D0769077

LOW-POWER CMOS WIRELESS COMMUNICATIONS
A Wideband CDMA System Design

NATIONAL UNIVERSITY
LIBRARY   SAN DIEGO

LOW-POWER CMOS WIRELESS COMMUNICATIONS
A Wideband CDMA System Design

by

Samuel Sheng
Datapath Systems, Inc.

and

Robert Brodersen
University of California

KLUWER ACADEMIC PUBLISHERS
Boston / Dordrecht / London

Distributors for North America:
Kluwer Academic Publishers
101 Philip Drive
Assinippi Park
Norwell, Massachusetts 02061 USA

Distributors for all other countries:
Kluwer Academic Publishers Group
Distribution Centre
Post Office Box 322
3300 AH Dordrecht, THE NETHERLANDS

Library of Congress Cataloging-in-Publication Data

A C.I.P. Catalogue record for this book is available
from the Library of Congress.

Copyright © 1998 by Kluwer Academic Publishers

All rights reserved. No part of this publication may be reproduced, stored in a
retrieval system or transmitted in any form or by any means, mechanical, photo-
copying, recording, or otherwise, without the prior written permission of the
publisher, Kluwer Academic Publishers, 101 Philip Drive, Assinippi Park, Norwell,
Massachusetts 02061

Printed on acid-free paper.

Printed in the United States of America

Table of Contents

Preface and Acknowledgments

The material presented in this book is the result of broadband wireless systems research done at the University of California, Berkeley, representing the work of several Ph.D. and M.S. theses over the past seven years. The wireless development was motivated by a much larger collaborative effort known as the Infopad Project, which was centered on developing a mobile information terminal for multimedia content – a wireless "network computer", so to speak. The desire for mobility, combined with the need to support potentially hundreds of users simultaneously accessing full-motion digital video, demanded a wireless solution that was far lower power and higher data rate than could be provided by existing systems. That solution is the topic of this book: a case study of not only wireless systems design, but also the implementation of such a link, down to the analog and digital circuit design level.

In Chapter 2, an overview of the system itself is presented, beginning with a brief description of the indoor radio-frequency environment and its various impairments, followed by a discussion of multiple access strategies. It is shown that a design utilizing spread-spectrum, code-division multiple access is ideally suited for such a system, not only as a means to multiplex the various users, but also to relax hardware performance requirements and to ultimately minimize power consumption in the receiver. The actual system specification follows in Chapter 3, documenting design decisions made on carrier frequency, channel bandwidth, modulation, multiple access, and spread coding, along with verification by system-level simulation.

In the remainder of the book, the actual hardware designed to implement the link will be examined. Chapter 4 examines the base-station transmitter, including a custom integrated circuit that performs user multiplexing, spectrum spreading, and channel pulse-shaping. The analog RF transmitter section is described in Chapter 5, consisting of a semicustom board design to upconvert the signal for transmission. Chapters 6 and 7 describe the custom analog integrated circuits developed for the mobile receiver, consisting of a low-noise amplifier, quadrature homodyne demodulator that utilizes passband sampling to achieve frequency conversion, a multi-stage variable-gain amplifier, and an extremely high-speed analog-to-digital converter. Low power and high integration are emphasized here, in achieving a "single-chip" monolithic CMOS solution for the analog front-end – the tremendous impact of the use of code-division multiple access will be made apparent. Following this, Chapters 8 and 9 examine the digital aspects of the receiver - the custom spread-spectrum digital demodulator circuit required to perform data and timing recovery, and the low-power techniques employed in its implementation. Lastly, conclusions and a description of future directions are presented in Chapter 10.

ACKNOWLEDGMENTS

Obviously, a project such as this is not accomplished without a large cast of supporting players. It was funded in part by the Department of Defense Advanced Research Projects Agency, with additional funding from Intel, IBM, Motorola, Texas Instruments, Ericsson, National Semiconductor, Hewlett-Packard, and the Fannie and John Hertz Foundation. Likewise, the authors would like to thank the members of the RF research group for their innumerable contributions to the project: Bill Baringer, Lapoe Lynn, Ian O'Donnell, James Peroulas, Kevin Stone, Craig Teuscher, Dennis Yee, and Jennie Chen. Professors Paul Gray, Robert Meyer, and Jan Rabaey, as well as their respective research groups, were most helpful with their astute technical advice: the many conversations with their students and them helped shape this project to a great extent.

In the moral support department: Jennie Chen, Andy Burstein, Tom Burd, Tony Stratakos, Chris Rudell, Cormac Conroy, Greg Uehara, Tom Truman, Monte Mar, Lars Thon, Shankar Narayanaswamy, Paul Landman, Lisa Guerra, Renu Mehra, Tom Boot, Peggye Brown, Kevin Zimmerman, and Brian Richards could all be counted on in a crisis. And lastly, special thanks goes to Peggy Huang for having suffered through proofreading the entire text, in gory detail.

LOW-POWER CMOS WIRELESS COMMUNICATIONS
A Wideband CDMA System Design

1 Introduction

In recent years, the need for personal mobile communications - "anytime, anywhere" access to data and communication services - has become increasingly clear. Digital cellular telephony, such as the U.S. Digital PCS and the European GSM systems, has seen rapid acceptance and growth in the marketplace. Likewise, the tremendous impact of the Internet has resulted in a growing demand for a new kind of data access: mobile retrieval. No longer content with being restricted to the office or home to read electronic mail or search for information on the Web, users are now looking for means to access their information while on-the-go. A simple notebook computer by itself does not suffice; although the computing capability is mobile, the communications must become mobile as well. Services such as the Metricom wireless data network and cellular digital packet data are becoming increasingly commonplace – a testament to the need for ubiquitous information and communications access.

However, the development of radio-frequency wireless systems is still in its infancy.

The above statement may seem somewhat startling, in light of the fact that radio frequency (RF) systems have been under development for the better part of the last century. However, existing RF technologies have been severely limited in either bandwidth or in portability. Clearly, the demand for portable voiceband services has resulted in intense research efforts to improve performance and system capacity, primarily through digital coding and modulation. However, these systems focus exclusively on wide-area, narrowband communication, while emphasizing portability. On the other hand, existing broadband systems, such as broadcast analog television and point-to-point terrestrial microwave, are simply not designed to support hundreds or thousands of individual users, each with their own unique data stream. Simultaneously meeting the needs of high per-user bandwidth and mobility is the goal of next-generation RF systems. Achieving this will require careful consideration of both system and implementation issues, as well as a close study of the interaction between the two.

This book is devoted to examining one potential solution to this problem: in particular, it focuses on the issues behind the development of a high-bandwidth, silicon complimentary metal-oxide silicon (CMOS) low-power transceiver system for mobile RF wireless data communications. In the design of any RF communications system, three distinct factors must be considered: the propagation environment in question, the multiplexing and modulation of user data streams, and the complexity of hardware required to implement the desired link. None of these three can be allowed to dominate; for example, implementation issues place constraints on answering "How many users?" and "How much bandwidth?", whereas the choice of modulation and multiple access can have dramatic impact on the questions of "How many transistors?" or "How much power?". This coupling between system design and implementation is key to simultaneously achieving high bandwidth and low power, and will be emphasized throughout the text.

From an design standpoint, one key focus will be the use of silicon CMOS fabrication technologies throughout. The importance of a silicon CMOS implementation cannot be overlooked. The tremendous speed inherent in deep submicron CMOS technologies enables operation in frequency regimes traditionally occupied by gallium arsenide or silicon bipolar technologies, while preserving the low manufacturing costs and high levels of integration that have made CMOS the predominant semiconductor technology. Furthermore, since portability places severe constraints on the physical size and weight of the transceiver, CMOS also allows the possibility of a single-chip or two-chip solution: a far cry from the dozens of components found in existing RF hybrids. This integration has another benefit: it enable new possibilities at the system level. Parallel digital signal processing can be readily employed, as compared to the conventional all-analog superheterodyne systems of today. Likewise, digital coding techniques can be brought to bear to combat impairments in the channel, as well as overcome implementation limitations such as thermal and quantization noise in the receiver.

Obviously, one does not simply design a wireless communications system with no eventual application in mind. Figure 1.1 shows one possible application of a next-generation mobile wireless system: to enhance and supplant existing wired networks by providing tetherless access. Indoor operation is assumed, with only short distances between the transmitter and receiver; this allows deployment in critical areas such as office buildings, airports, and homes, while providing a relatively benign transmission channel (Chapter 2). The key target data rate criteria is to support, for every user, full-motion digital video, which demands a minimum per-user data rate of 2 Mbps including error correction and coding. Lastly, an asymmetric data rate system is assumed, in that the uplink data from the user is at a much lower rate than the downlink data going to the user. If one considers the way the Web accessed today, most of the data coming from the user consists of keystrokes or voice audio, which can be supported at data rates as low as 32 to

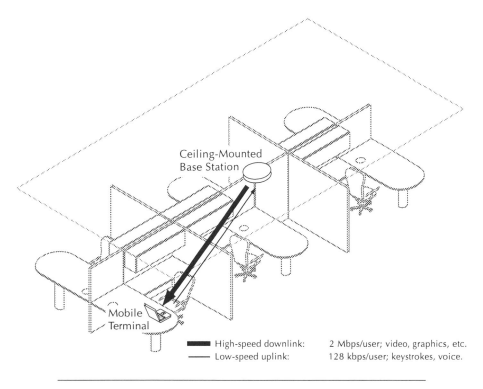

Figure 1.1: Potential application of a next-generation RF transceiver

128 kbps/user.[1] Since the low-rate uplink can be handled by any number of existing implementations [Cylink96,Prox94], it is the high-rate downlink – transmission from a fixed base station to the mobile receiver – which is of greatest interest.

Even with the constraints of asymmetric data rates and indoor operation, the above still presents a daunting design challenge. Due to the high per-user data rate requirements, carrier frequencies must be placed in or above the low microwave bands, owing to spectrum congestion at frequencies below 1 GHz. Likewise, a large number of users needs to be accommodated by the system simultaneously, densely packed into a small physical area; continuous network access by 1 user every 4 to 10 square meters is typical for an office environment. Ostensibly, a microcellular or even picocellular reuse system must be employed to come remotely close to the required capacity [Lee89]; however, even with cellu-

---

1. In fact, there are several wired precedents for such an asymmetric configuration: the asymmetric digital subscriber loop [ADSL95], cable modems, and the current 56 kbps voiceband modems.

lar reuse and a limited number of users per base station, the required aggregate downlink channel data rates will be in the tens to hundreds of megabits per second. Because of this, the most difficult implementation aspect is the development of a low-power, silicon CMOS receiver for the mobile, a receiver capable of handling hundreds of megabits per second, while consuming only milliwatts of power.

To summarize:

- The analog circuitry must perform reliably at variable carrier frequencies above 1 GHz, while supporting a sustained per-user data rate in excess of 2 Mbps.

- The system must support simultaneous access by a large number of users, within a small physical area.

- The mobile transceiver cannot consume an excessive amount of power, as the portable terminal must be powered by batteries, nor can it be excessively large.

Existing commercial solutions fall far short of this in terms of both bandwidth and power consumption. To truly achieve a broadband, mobile wireless system, a thousandfold improvement in the power-throughput ratio over existing hardware is needed - for the receiver, 1 milliwatt per Mbps of received data rate, while supporting a raw channel capacity in excess of 100 Mbps.

The remainder of this book is entirely focused on the development of techniques by which such a thousandfold improvement can be achieved. The solution to achieving this can be summarized quite succinctly: leverage off of the parallel digital processing techniques achievable by today's technologies. In particular, system- level designs that inherently favor complex digital signal processing, minimizing analog complexity and power consumption, are both necessary and desirable. All aspects of such a wireless system design will be discussed: ranging from low-level analog design, where a novel receiver structure using passband sampling is presented, up to the multiple access schemes utilized and its impact on the transceiver system.

In Figure 1.2, a high-level system block diagram of the wireless broadband downlink is shown. Each chapter of this book is devoted to examining each of the major blocks of the overall system. Chapter 2 will examine the characteristics of gigahertz-band RF transmission indoors, as well as a short overview of modulation and multiple-access strategies. Following this, Chapter 3 focuses on the specific link description, including a discussion of system capacity, carrier frequency, transmit power, modulation, and coding. The base-station transmitter will be covered in Chapters 4 and 5, with one chapter devoted to the baseband modulator, and the other devoted to the RF upconversion circuitry. Lastly, the implementation aspects of the receiver are examined in Chapters 6 through 9: RF front-end circuitry, the analog baseband circuitry, and the baseband receiver digital signal processing. Throughout, the interrelationship between the implementation and the system design is emphasized, especially with regard to the use of silicon CMOS and the requirements of low power design.

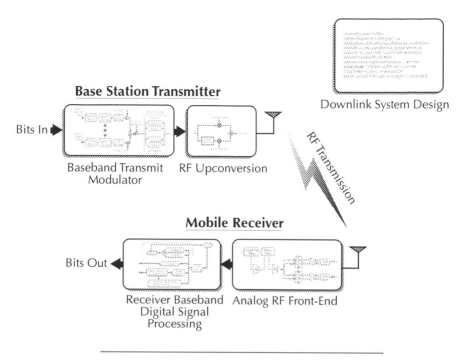

Figure 1.2: Broadband downlink system diagram

REFERENCES

[ADSL95]   *ADSL Metallic Interface Standard.* ANSI Document T1E1.4/95-007R2/T1.413.
                American National Standards Institute, 1995.

[Cylink96]   Airlink 19.2-512 Radio, Cylink Product Catalog. Cylink, Inc.,
                Santa Clara, CA, 1996.

[Prox94]   RangeLAN2 Wireless Modem Product, Press Release. Proxim, Inc.,
                Mountain View, CA, March 30, 1994.

[Lee89]   W. C-Y Lee. *Mobile Cellular Telecommunications Systems.* New York:
                McGraw-Hill Book Co., 1989

[Symb95]   Symbol Technologies, Inc. *Spectrum 24 Wireless Modem Product Press Release.* Holtsville,
                NY, June 12, 1995.

2 Modulation, Multiple Access, and How Radio Waves Behave Indoors

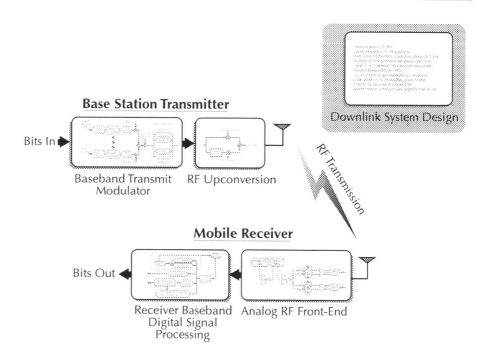

To begin, the system design of the high-speed downlink will be discussed in this chapter and the next. Given that an in-building solution is desired, a description of the propagation characteristics of the indoor channel will first be presented. This is followed by a short exposition of the available digital modulation and multiple-access strategies, describing the advantages and disadvantages of each. In Chapter 3, the actual system link specification will be given – in terms of number of users per base-station, transmit bandwidth, modulation strategies – and a discussion of how this specification was devised.

2.1. Cellular Reuse and the Indoor Propagation Channel

Since the system must support independent full-motion digital video streams to each user, spectrum usage becomes of great concern, requiring a per-user data rate of 2 Mbps. Also, this data rate is clearly not needed on a continuous basis; when regular computation tasks are being performed, such as using a word processor or a spreadsheet, the screen changes only slightly on a frame-by-frame basis and over only a small region, usually on the order of a single character or a few pixels. Hence, it is easily possible that the peak data rate required by a user is much larger than the overall time-average data rate, and minimizing overall system bandwidth consumption while supporting a large number of users accessing data simultaneously is of paramount importance. One technique, applied at the system level, is to utilize cellular networking to achieve spatial frequency reuse.

The advantages in improved spectral efficiency afforded by cellular systems have long been known; having been employed extensively in present-day analog mobile radiotelephony, the large-scale cells utilized only exploit these advantages to a limited extent. By scaling down cell sizes, tremendous increases in spectral efficiency can be achieved. In this section, an analysis of these gains is presented, along with a description of the cellular concept as realized in an indoor environment. A statistical characterization of the indoor picocellular transmission environment and the corresponding model are then described.

2.1.1. Cellular Systems

As shown in Figure 2.1, a simple cellular scheme consists of dividing the entire service area for the personal communication system into "cells" of radius R, with a single base station serving all mobile users within that cell; each cell utilizes its own distinct set of system resources, different from its surrounding neighbors. As an example of such resource allocation, conventional analog cellular, being a frequency division multiple-access system, allocates different frequency bands to each cell. As will be shown later, there is a clear correlation between the method of resource allocation and the multiple access strategy: in time-division multiple access, each cell could be assigned different time slots, and in code-division multiple-access, each cell could be assigned different codes.

As users move from cell to cell, their transactions with the network are "handed off" from base station to base station, reconfiguring the network dynamically as the need arises. Such reconfiguration is thus accomplished locally, since the system itself is responsible for determining where and when handoffs occur. Clearly, the complexity involved with cellular network control is much greater than that required for a classical "umbrella" scheme, with one base antenna for the entire service area, and much of the intelligence of the network must be deployed within the base station and the mobile transceivers.

The key advantage to cellular systems is that they allow the network to achieve spatial multiple access of the users. If two cells are separated by sufficient distance, each can use the same channel resources at the same time without resulting in disastrous cochannel interference. Figure 2.2 shows several classical reuse patterns [Lee89]; such patterns are typ-

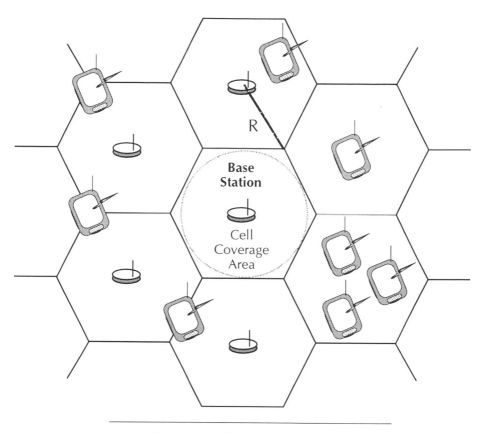

Figure 2.1: Cellular transmission system

ically characterized by a reuse factor K, which represents the number of distinct sets of resources (either frequencies, or time slots, or codes, representing several possible user multiplexing strategies) that need to be used to cover the entire service area.

From the point of view of spectrum usage, each user effectively consumes only B/N Hz of bandwidth, where B is the raw physical bandwidth needed to support transmission within a single cell, and N is the total number of users in the system. Hence, cellular systems are said to be spectrally efficient. Clearly, minimizing the distance D between cells utilizing the same frequency yields the greatest frequency reuse, since the number of cells that can use the same frequency band in the service area is maximized, and hence the greatest gain in spectral efficiency is achieved. This frequency reuse distance is geometrically related to K and R by $D=R\sqrt{3k}$. In terms of system capacity, supposing that the total service area serving the N users is given by $\pi\rho^2$, then the number of cells is given by $(\rho/R)^2$ and hence the total bandwidth for the system has been reduced by a factor equal

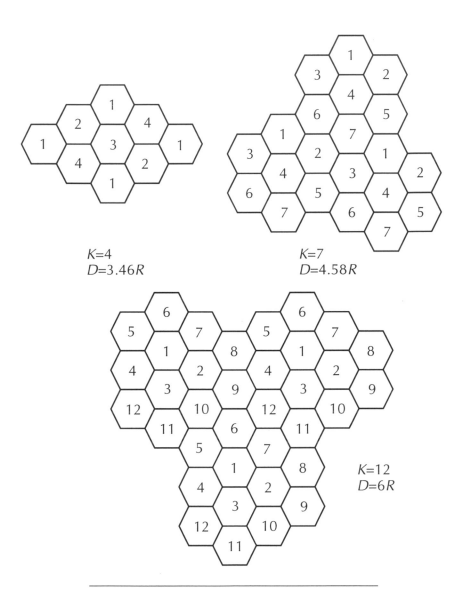

Figure 2.2: Typical cellular reuse patterns

to $(R/\rho)^2K$ relative to an equivalent non-cellular system. As shown in Figure 2.3, the normalized bandwidth required by a cellular system is plotted as a function of R and K; since it is quadratically dependent on R, it is of greatest benefit that the cell radius be reduced. Clearly, when $R=\rho$ and $K=1$, the original "umbrella" scheme is yielded.

Ostensibly, the ultimate limits on minimizing R (and hence D) lie in how much cochannel interference the system can tolerate, and the required complexity in network control. An important fact is that the level of cell-cell cochannel interference is independent of the scaling of R, since the transmit power in each cell scales with R and hence the relative interference stays constant. Thus, cochannel interference is only a function of K. Conversely, complexity in network control is only a function of R, since more handoffs will necessarily occur as R decreases. Both K and R are highly dependent on the transmission environment; however, these massive gains in spectral efficiency clearly will be needed to simultaneously support the desired high-speed and high-bandwidth requirements.

2.1.2. The Indoor Picocellular Environment

Within an indoor environment, it is no longer feasible to have only a single network transceiver station serving all of the terminals in the building. Due to the 5 to 15 dB attenuation through walls in the low gigahertz bands [Seidel91], the total microwave output power from all of the transmitters would have to be inordinately (and dangerously) high. However, this attenuation can be taken advantage of by a cellular network. Each room naturally becomes its own cell; likewise, the cellular scheme moves into three dimensions, since the floors also provide RF isolation. As the cells are now extremely small, on the order of a two to five meters, this cellular strategy is commonly referred to as microcellular or picocellular networking. Hence, R is usually dictated by the size of the room, and K can be as low as 3 to 4, depending on how much attenuation is provided by the walls. If K is increased to 6 or 7, the assumption that cochannel interference is negligible becomes reasonable for most indoor office environments.

In light of the above considerations, the total amount of spectrum that will be consumed to provide the outlined services can now be addressed. After examination of the user density in a typical office environment, such as those found in modern buildings with open-area soft-partition cubicles, cells of 5 meter radius typically contain twelve to sixteen active users. A 2 Mbps data rate for full motion video (our worst-case requirement), using a linear DQPSK (differential quadrature phase-shift keying) modulation scheme and using design parameters from existing systems [Feher87, Raith91], would require a transmission bandwidth of approximately 1.6 MHz per user using a 30% excess bandwidth raised-cosine pulse shape and a 25% loss for packetization, synchronization, and other overhead. Assuming that, of the 16 users in the cell, half of them are demanding the complete 2 Mbps data rate for full-motion digital video with the remainder utilizing 256 kbps [1] each for lower data rate applications, a picocellular system with K= 7 would utilize approximately 100 MHz of bandwidth. Although 100 MHz is a considerable amount of spectrum, this is amortized over large numbers of people using this spectrum simultaneously within multiple buildings. Considering that the bandwidth of 100 MHz is designed to support full motion video and other multimedia network services for all users, this allocation of spectrum is not unreasonable given the level of service provided by the system, especially when compared to the spectrum allotted for existing systems such as NTSC television.

2.1.3. Statistical Characterization and Multipath Propagation

Unfortunately, the indoor environment also presents several transmission difficulties, the dominant one being the numerous reflections of the radio signal off walls, furniture, and even people. A large number of paths exist between transmitter and receiver; the received

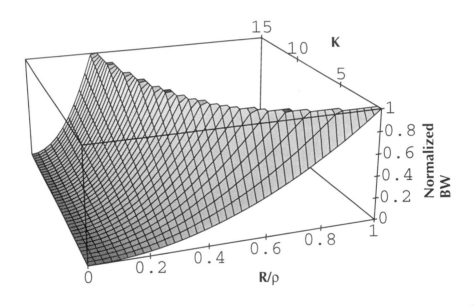

Figure 2.3: Normalized bandwidth of cellular systems

Normalized with respect to the non-cellular case, assuming a
constant number of users and data rate.

_____

1. 256 kbps is reflective of data rates used in existing X-terminals; this value may be considerably lower depending on the level of user activity.

signal is thus corrupted by multipath distortion and inter-symbol interference, since these reflections can be significant over the span of several symbols. Also, in a picocellular network, it is likely that there is no line-of-sight propagation path due to shadowing, so transmission becomes critically dependent on the reflected waves. All of these effects vary slowly with time; even with the remote unit motionless during operation, the motion of people can cause significant variations in the environment. The critical statistical parameters to be determined are: the number of paths, the excess time delay for a single path, the path loss, and the time variation of the received signal, for a cell size of 2 to 10 meters. Using measured data collated from several sources [Seidel89, Seidel91, Ohrvi89], a characterization of the indoor propagation environment can be determined.

First, the path loss in a multipath situation must be considered. It has been found experimentally [Ohrvi89] that the total received power at a particular distance d of a multipath profile can be modeled as a log-normal (normal in dB) distribution about a local mean path-loss law given by d^{-n}, valid for d approximately larger than λ, where λ is the carrier wavelength at 1 GHz = 0.3 m. Values of n ranging from 1.5 to 6 have been reported, although typical values range from 1.5 to 3. Figure 2.4 shows a measured path loss characteristic [Ohrvi89], with the best-fit curve varying as $d^{-2.6}$. It has also been found that, for a single path component, the received power has a log-normal distribution about an exponential path-loss law $d^{-n(\tau)}$, where τ is the excess delay time defined as $t_{arr} - t_0$ and t_0 is the delay of the line-of-sight path, with an obstructed topology exhibiting greater attenuation than a line-of-sight one as a result of the extra path distances incurred by shadowing. Generally, the standard deviation of the log-normal distribution is insensitive to τ, and is approximately 4 dB for line-of-sight topologies, and 5 dB for obstructed ones.

The problem of the statistics of the number of paths and their arrival times at the receiver also needs to be considered. Clearly, receiver sensitivity here is critical, since more paths can be resolved by a receiver with higher sensitivity. For low-sensitivity receivers the number of path has been measured to be approximately Poisson, with a mean of 4.4 paths and standard deviation of 2.1, for a receiver threshold of 30 dB below a 10λ received power reference. As receiver sensitivity increases, a Poisson distribution becomes a poorer model for the number of paths, and a normal distribution with a mean of 22.4 paths and standard deviation of 8.6 paths provides a good statistical model for a 48 dB threshold. The other path-dependent factor is the distribution of arrival times; having determined the behaviour of the number of paths, the probability of a path having an excess arrival time τ needs to be characterized. From several measurements [Seidel89, Saleh87], the excess arrival times follow an exponential distribution, with a mean time of 30 nsec for transmitter-receiver separations below 10 meters.

One important parameter related to the path distribution is the total delay spread, or the total time between the first arrival and the last resolvable arrival due to multipath. The delay spread characterizes the signaling rate below which the fading will be "flat"; in other words, if the time for each transmitted bit is greater than the delay spread, there will be no intersymbol interference, since all of the reflections will have died away before the next bit occurs. In the frequency domain, this quantity is referred to as the coherence

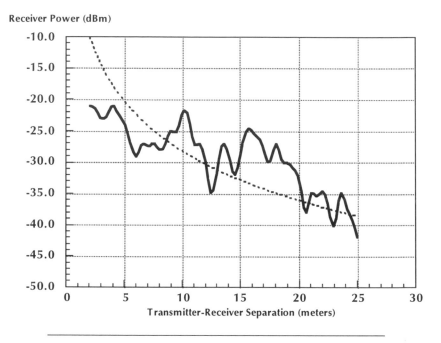

Figure 2.4: Measured path loss characteristic
(from [Ohrvi89])

bandwidth, which can be intuitively viewed as the interval (in Hertz) between consecutive fading nulls in the channel impulse response. The relationship between delay spread T_{del} and coherence bandwidth C_{BW} is that C_{BW} is approximately equal to $1/T_{del}$. For indoor channels, at a sensitivity of 48 dB below a 10λ threshold, T_{del} is approximately 30 nsec, corresponding to a coherence bandwidth of 33 MHz. As will be shown, it is desirable to either transmit at a data rate far below the coherence bandwidth (hence resulting in no intersymbol interference and flat fading), or at a data rate far above the coherence bandwidth, in the case of spread-spectrum.

Unfortunately, the difficulties presented by the multipath environment are further compounded by the fact that the environment is also time-varying, resulting in fading. Fading occurs when the environment varies such that destructive interference occurs at the receiver and causes much more attenuation than predicted by path loss alone. Since it is assumed that the remote device is relatively stationary, the typical Rayleigh-distributed fading patterns found in analog mobile phone environments are not encountered. Instead, the primary cause of time-variation is the movement of people in the propagation path. For only a few people, the distribution has been found to be Rician more than Rayleigh; however, as the number of persons increases, the distribution becomes increasingly

Rayleigh. From the measurements taken, a typical fluctuation results in a signal deviation 10% of the time in excess of 4 dB away from the median, and a deviation in excess of 8 dB 1% of the time [Ohrvi89].

Lastly, noise effects within the channel must be considered. From measurements taken of the 1-2 GHz frequency band, the additive white Gaussian noise present on the channel is essentially at the thermal noise floor (-110 dBm at 300°K over 1 MHz); hence, as far as Gaussian noise is concerned, the noise contribution is dominated by the noise at the input of the receiver. However, the low-microwave band also exhibits significant impulsive noise, resulting from RF interference from such sources as computer systems, power microwave cavities in cooking ovens, and common relay switches. It is extremely difficult to characterize these sources, since the presence and characteristics of such generators vary so widely, even within the same building. From some recent measurements, it has been shown that the average noise factor above the thermal noise floor for impulsive sources can be as high as 50 dB [Black91].

2.1.4. Channel Modeling

For multipath transmission, an early time-varying, wideband model proposed by Turin is given as:

$$\sum_{k=0}^{n} a_k e^{jq_k} s(t - t_k) + n(t) \qquad .$$

where s(t) is the complex-valued, lowpass representation of the transmitted pulse, n(t) is the noise component, q_k is the carrier phase shift, and a_k are the amplitudes of each arrival at time t_k in the multipath profile. Since along each propagation path, the physical media is essentially "ideal" (i.e., represented by a simple attenuation constant), the impulse response of such a channel can thus be characterized by an expression of the form:

$$h(t) = \sum_{k=0}^{n} a_k \delta(t - t_k) e^{jq_k}$$

with $\delta(t)$ equal to the Dirac delta function. By using the statistical data described above, a model of this impulse response suitable for simulation can be developed.

The path strength coefficients a_k are set to the local mean value given by d^{-n}, where n=2.6, multiplied by the variation that is normally distributed in dB. Since of primary interest is a simulation model suitable for verifying the functionality and performance of the indoor transceiver downlink, the time-variation of the channel is sufficiently slow that we can assume it is constant over the simulation time. For our model the path attenuations are thus fixed at the beginning of the simulation and remain invariant. Likewise, the number of paths and individual path delays are treated similarly. At the start of the simulation, the number of paths is randomly determined, which is normally distributed as described in the previous section. For each path, an exponentially distributed random value is assigned, corresponding to the path delay.

If long-term error performance simulations are required, such as large Monte-Carlo estimations of the bit-error rate, the model can be adapted for such purposes, by simply incorporating time variability into the model. The only consideration that needs to be made is that over time, the amplitude and phase of each multipath component must be continuously related to the original randomly generated profile – the profile cannot be "randomly" updated. Statistically speaking, the resulting fading behaves as a Rician process; by folding in a time-series model, time variability can be accounted for.

To give an idea of how severe the fading environment can be[1], the impulse responses (both in time and frequency) of two such randomly generated channels are shown in Figure 2.5; they are quite similar to impulse responses measured in the field. The magnitude frequency response of the channel, $H(j\omega)$, is plotted in a 300 MHz bandwidth in the neighborhood of 1 GHz. In comparing the two, the dynamic range due to indoor fading can easily exceed 30 dB, consistent with measured results. Unless such nulls are combatted, signal degradation and possibly complete link breakdown will result, since the signal-to-noise ratio has been compromised by the null.

As an important application of this model, the effects of the multipath on an actual transmitted signal will be simulated, by generating an eye diagram of the received signal. Using a simple 4-PSK QAM modulation scheme with white input data, the effect of the multipath distortion is examined for a data throughput of 1 Mbps at a carrier of 1 GHz. As described above, the baseband Multipath block is used heavily in this type of simulation. A 50% excess-bandwidth raised-cosine filter response is assumed for the combined transmit and receive bandlimiting filters, with the filter partitioned equally between the transmitter and the receiver, i.e., the transmit filter has a frequency response equal to $\sqrt{H(j\omega)}$, where $H(j\omega)$ is the transfer function of the raised-cosine lowpass filter, and similarly for the receive filter. Also, the amplitude of the transmitted signal is normalized to unity, since only the relative loss is of real interest.

In Figure 2.6, the ideal (no channel) eye diagram of the in-phase received signal is shown after a transmission of 100 symbols; as expected, the "eyes" are fully open, and the signal value there is precisely contained in the set {-1,0,1}, indicating zero distortion and perfect recoverability. The waveform in between the sampling points is the response of the raised-cosine filter; this also verifies that the filter responses were designed correctly, as the Nyquist zero-forcing criterion is clearly achieved. In Figure 2.7, the channel is now inserted between the transmitter and receiver in the system, and the resulting eye diagram is shown. Demonstrating the key difficulty in transmitting within such an environment,

---

1. It should be mentioned here that the simulation and system verification were accomplished using the U.C. Berkeley Ptolemy simulator [Buck93]; details of the actual simulation code are contained in [Sheng91].

Simulated channel, in-phase primary arrivals at 1 GHz (strong peak)

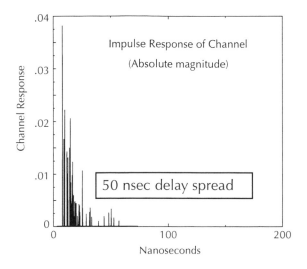

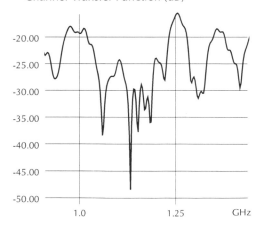

Figure 2.5: Simulated random channels (this page and next)

(this page) Strongly in-phase arrivals
Narrowband dynamic range > 30 dB
Delay spread ~ 50 nsec

Simulated channel, out-of-phase primary arrivals at 1 GHz (strong null)

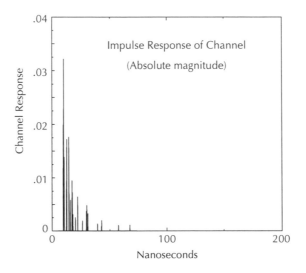

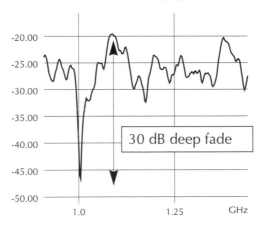

Figure 2.5 (continued): Simulated random channels

Strongly out-of-phase arrivals
Narrowband dynamic range > 30 dB
Delay spread ~ 50 nsec

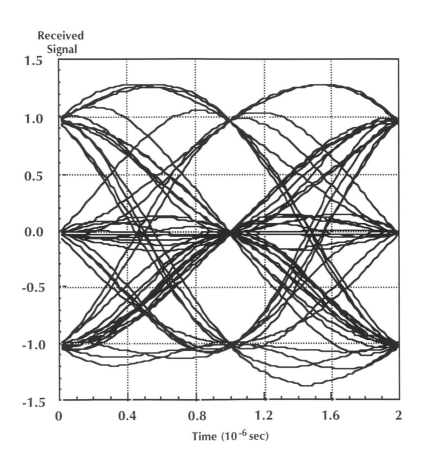

Figure 2.7: Simulated QPSK eye diagram, no channel ISI

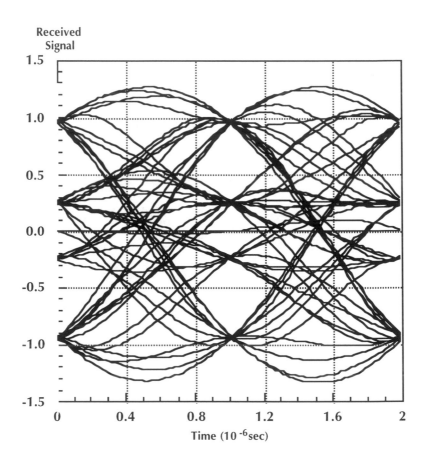

Figure 2.8: Simulated QPSK eye diagram, with multipath channel

the eye openings are now relatively closed due to the multipath distortion. The "zero" level has clearly split into two; this is to be expected, since when the in-phase symbol has value zero, the quadrature signal is one. Due to the random phase shift caused by the channel, the in-phase signal is corrupted by crossover from the quadrature phase.

2.2. Modulation Techniques

When considering the design of the baseband signal modulation, it is important to note that this discussion is essentially restricted to the digital domain, and hence complex filtering and signal generation are not a problem. For example, simple frequency-shift keying, which uses different frequency tones to encode the data, traditionally utilizes a low-frequency analog voltage-controlled oscillator to generate the tones, and then modulates this signal up to the passband. However, settling time, temperature stability, phase noise, and other nonidealities in the baseband oscillator often complicate the system design. Today, a direct-digital frequency synthesizer can generate the baseband signals [Nichol91], without any of the difficulties encountered with analog designs. The use of digital baseband processing allows one to limit all of the analog hardware to the passband conversion circuitry, and opens the possibility of simplifying the analog hardware at the expense of more complicated baseband processing. These concepts will be explored more fully in subsequent chapters; at this point, it is important to realize the complexity in modulation synthesis is no longer as great of a concern as it once was, given ready access to digital signal processing. However, the hardware in the receiver - required to recover the modulated signal - must still be minimized, given the need for low power and high performance.

2.2.1. Quadrature Amplitude Modulation

Quadrature amplitude modulation (QAM) is simply the digital version of classical analog AM, using modulation on both the inphase and quadrature signals to achieve the same spectral efficiency as single-sideband modulation [Lee88]. Essentially, a sinusoidal carrier waveform is linearly modulated by a bandlimited signal of the form:

$$m(t) = \sum_{k=0}^{\infty} A_{kT}\, g(t - kT)$$

where $\{A_{kT}\}$ are the complex-valued transmit symbols corresponding to the user data, and $g(t)$ is a bandlimited baseband pulse. This results in a transmitted signal $s(t)$ equal to:

$$s(t) = 2\mathbf{Re}\left\{ e^{j\omega_c t}\, m(t) \right\} = \sum_{k=0}^{\infty} 2\mathbf{Re}\left\{ (A_{kT} g(t - kT))\, e^{j\omega_c t} \right\}$$

The $\{A_{kT}\}$ data symbols are chosen from a symbol constellation, as shown in Figure 2.8; each group of bits to be transmitted is encoded into the complex-valued symbol shown. Clearly, the larger the constellation, the more bits are transmitted per symbol, and hence the required transmission bandwidth is reduced. However, since the distance between constellation points is reduced as the constellation size increases, the probability that a symbol error occurs also increases. Under additive white Gaussian noise conditions, the probability of a bit error for a QAM signal is approximately given by:

$$P_{error} = \left(\frac{\eta}{M\sqrt{2\pi}}\right)\int_{q}^{\infty} \bar{e}^{(x/2)^2} \, dx = \left(\frac{\eta}{M}\right)F_{err}(q) \leq \left(\frac{\eta}{M\sqrt{2\pi}}\right)\exp(-q^2/2)$$

where $q=(d_{min}/2\sigma)$, d_{min} is the minimum distance between two nearest-neighbor constellation points, σ is the noise power, M is the number of bits per symbol, and η is a degeneracy factor equal to the number of nearest neighbors at the minimum distance.

The primary advantage of QAM is that it is known to be spectrally efficient, which is ideal for a personal communications system with its severe bandwidth requirements; also, it is conceptually simple to equalize out the effects of the channel since the use of adaptive linear equalization is possible. However, the receiver hardware complexity immediately comes into question, since coherent demodulation has traditionally been utilized requiring phase- and carrier-recovery circuitry. For these high data rates at such high carrier frequencies, essential components such as rejection filtering, carrier recovery, automatic gain control, and voltage-controlled local oscillators become complex and expensive to implement. If QAM is to be employed, methods of minimizing the receiver complexity must be developed if the spectral efficiency of QAM is to be exploited; for example, the use of differential phase encoding, instead of absolute phase, obviates the need to perform coherent data recovery.

2.2.2. Continuous-Phase Modulation

Continuous-phase modulation (CPM) techniques present an interesting class of modulation schemes that have the desirable property that they can be incoherently detected with minimal hardware complexity. They are characterized by a constant amplitude envelope, with all of the information carried in the phase of the signal alone. The simplest form of CPM is FSK (frequency-shift keying), in which two separate frequencies are used to distinguish "0" and "1" data bits [Lee88]. The general form of the transmitted signal is given by [Ekel87]:

$$(s(t) = \sqrt{2(E_s R)}\cos(\omega_c t + \varphi(t, \alpha))), \quad \varphi(t, \alpha) = 2\pi h \sum_{k=0}^{\infty} A_{kT} q(t - kT)$$

where $q(t) = \int_{-\infty}^{t} g(\tau) \, d\tau$, E_s is the energy per symbol, and A_{kT} are the transmitted bits.

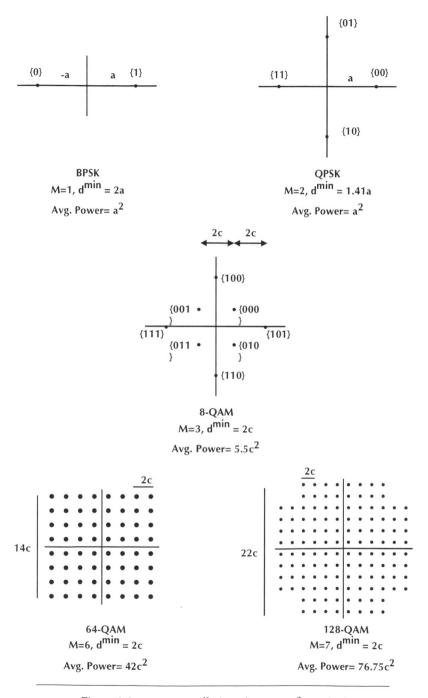

Figure 2.9: QAM constellations (cross-configuration)

Since this is a constant envelope modulation scheme, CPM is power efficient at the transmitter, since a class-C nonlinear power amplifier can be used. Likewise, it is immune to channel amplitude nonlinearities; for some schemes the receiver can simply hard-limit the signal without any loss of information or signal degradation, obviating the need for high-precision automatic gain control as needed by linear modulation schemes. Another important advantage is that most CPM schemes can be demodulated incoherently, in that the absolute phase of the received carrier need not be known for data recovery; for example, to demodulate FSK, a simple bandpass filter bank can be used to detect the various frequency tones, thus recovering the data without need for complex phase recovery.

Conceptually, CPM schemes are the digital equivalent to analog FM, with the digital signal first being low-pass filtered before frequency modulation, as shown in Figure 2.9. The name CPM comes from the requirement that the phase be continuous at all points, which implies that no "jumps" in phase are allowed, as in simple FSK. This continuity requirement improves the spectral efficiency significantly, and it is the bandwidth of the low-pass filter which determines the overall spectrum consumption of the system. A simple extension of FSK that meets this continuity requirement is minimum-shift keying (MSK) [ProakB83]; it does so by a careful choice of harmonically-related signaling frequencies and signal polarities. However, it is not spectrally efficient; one related "spectrally efficient" CPM scheme is GMSK, or Gaussian Minimum-Shift Keying[1] [Ekel87, Murota81]. The low-pass filter is specified to have a Gaussian response, and it has been shown that to maximize the spectral efficiency the bandwidth of the Gaussian filter should be set to approximately 0.25R, where R is the bit rate. To compare with simple MSK, the 99% power containment bandwidth for GMSK can be shown to be 0.86R [Murota81], as opposed to 1.2R for MSK. However, the spectral efficiency of CPM is not as good as that of QAM – given the severe bandwidth constraints in the desired link, this makes CPM intrinsically unattractive as a modulation format for short-range, high-bandwidth in-building transmission. This is in direct contrast to digital cellular voiceband systems, where high

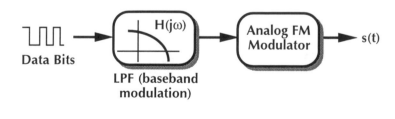

Figure 2.10: General CPM modulator

---

1. GMSK has been of the object of significant interest recently, since it is the modulation scheme that has been specified for the widely-used GSM digital voice cellular system.

output power and power amplifier efficiency are absolutely critical, and CPM schemes are intrinsically favored. The short transmit distances ameliorates the required output transmit power tremendously, and thus minimizing the impact of power amplifier efficiency; spectral efficiency becomes the overriding concern.

2.3. Multiple Access Strategies

Multiple-access techniques are methods by which many users can access the system simultaneously, where users are separated from one other by allocating a small slice of the available resources to each user. For example, when cellular technology is employed, a form of spatial multiple access is achieved, in which two users separated by a sufficient distance can independently use the same frequencies at the same time. Another classical scheme is frequency-division multiple access (FDMA), where users are separated from each other by assigning a distinct frequency band to each. Hence, two sets of independent data can be transmitted at the same time at the same location, without resulting in interference between the two users. Below, a short discussion is presented for three other important schemes: time-division multiple access (TDMA), direct-sequence spread-spectrum (code-division) multiple access (DS/SSMA, or CDMA), and frequency-hopped spread-spectrum multiple access (FH/SSMA), which is a variant of FDMA.

Of importance is the fact that FDMA, TDMA, and direct-sequence CDMA can all be viewed as partitionings of the available signal space into N subsets, where N is the number of users that are to be multiplexed. Optimally, N is equal to the dimensionality of the signal space, which has been shown to be of dimension 2BT, where B is the total available bandwidth for transmission, and T is the amount of time that is available for the N users to all transmit one symbol. For example, in FDMA each user of the N users transmits in a bandwidth of B/N, over the complete time interval T. As discussed below, for TDMA each user transmits in separate bursts of time of length T/N, over the complete bandwidth B, and for direct-sequence CDMA each user is assigned a one-dimensional subspace derived from one of 2BT orthogonal basis vectors for the signal space. The important realization here is that no one scheme is inherently superior to any other; it is only after considering the transmission environment and the required performance that advantages and disadvantages can be discussed.

2.3.1. Time-Division Multiple Access

In TDMA, the users are separated on a common communications medium by ensuring that they simply transmit at different times. Perhaps the most obvious form of multiple access, it is surprising simple in concept, and variants have long been used in satellite communications and wired local area networks such as Ethernet. Transmitted bursts of "low data rate" user information is multiplexed and interleaved into time-slices on a "high data rate" communications channel, as shown in Figure 2.12. Received TDMA data is put through a demultiplexor to reverse the interleaving process, extracting only the time slices that belong to that user. The time slices of user data are typically placed into frames, where each frame contains overhead acquisition information for timing and carri-

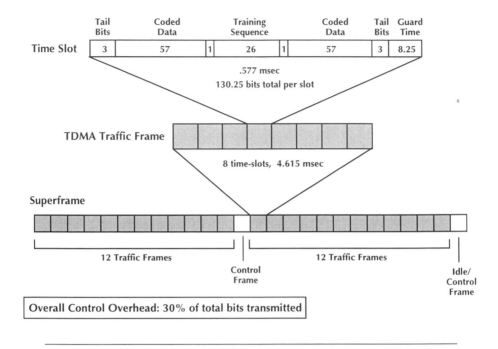

Figure 2.11: TDMA framing/control strategy
(for the GSM digital voice cellular system)

er recovery (framing bits) in addition to the data itself. From this, a considerable amount of overhead is clearly expended in reacquiring synchronization and training the adaptive equalizers from frame-to-frame; for the system of Figure 2.12, 1 in 3 of the bits transmitted is used only for control purposes.

TDMA multiplexing has been heavily favored for the past several years for use in second- and third- generation microcellular digital radiotelephony systems such as the European GSM project [Raith91], mainly because of its simplicity and ease of implementation. However, problems have already been encountered with the use of TDMA on microcellular channels; surmounting the time-varying channel multipath distortion and fading necessitates the use of an adaptive decision-feedback equalizer (DFE) [Sven90].

The usual design rule is that the number of taps in the equalizer structure should provide a time-delay span approximately equal to the length of the channel impulse response that needs to be equalized. For low-data rate (100 kbps TDMA burst rate, 32 kbps average user rate) voice telephony applications, this implies that a 3-forward, 7-reverse tap adaptive equalizer[1] should be sufficient, where the number of taps is the size of the forward filter and feedback filter respectively [Proak83]. Interestingly, for the indoor picocellular channel that we are considering, a 3-7 tap structure would also suffice in spite of the higher data rate. From the analysis in Section 2.1.2, a 100 MHz burst data rate is needed to support the necessary throughput to all users in the downlink, a thousand times faster than the typical voiceband cellular channel. However, the cell size has also shrunk a factor of 1000, with a commensurate compression in delay spread, thus implying that the equalizer structure will remain about the same complexity. The difference is how fast the receiver digital circuitry has to operate: the equalizer needs to function at the full 100 MHz bandwidth, which is not trivial even with today's technology. Furthermore, to enable these equalizers to converge sufficiently quickly, use of the modified least-mean squares or the Kalman filter algorithm is needed. These algorithms are complex, and tend to be computationally intensive in the receiver. Given the complexity and difficulties that have been encountered in forcing even small adaptive decision feedback equalizers (DFE's) to converge quickly, it will be difficult to design an adaptive DFE that will meet the performance requirements of a downlink. The only mitigating factor is that the multipath profile is changing slowly with respect to time, allowing slower convergence rates, as opposed to the typical fast-fading pattern encountered with operation of cellular transceivers in a moving vehicle.

2.3.2. Frequency-Hopped Spread-Spectrum Multiple Access

A variant of FDMA, frequency-hopped spread spectrum allocates each user a small slice of the total available spectrum, where each slice is equal to the user's data bandwidth. Unlike pure FDMA, the carrier frequency at which each user is transmitting is changing with time, hence the name frequency-hopped. The spectrum is "spread" in the sense that the user ranges over the entire transmit band due to the hopping, but only occupies a small amount at any given instant.

Although it would seem that FH/SSMA would be subject to the same disadvantages that TDMA has, the transmitted bit rate for FH/SSMA is the same as the data rate, unlike the case of TDMA. Hence, the fading is narrowband - the channel transfer profile is approximately flat over the bandwidth, and hence much less equalization is needed. Relative to the symbol rate, the time-span of the multipath is much shorter than that of TDMA, since we do not need to transmit in high-frequency bursts. In Figure 2.13, the simulated deep-fade transfer function from Figure 2.5 is plotted again, except with the bandwidth occupancy of the TDMA and FH/SSMA shown; the narrowbanding effect is evident. Of course, this narrowbanding advantage would also be realized with pure FDMA - no mention of the necessity of hopping has been made yet. Also shown in Figure 2.13 is an example hopping pattern for FH/SSMA - on hop number 2, it drops directly into a 30 dB fading null. In all likelihood, with such a deep null, the narrowband signal will be completely engulfed by noise, resulting in link breakdown. It is the fact that it will quickly hop back out of the null that makes FH/SSMA a viable multiple access scheme at all.

1. These data values are for the GSM digital voice cellular system.

The major disadvantage of FH/SSMA arises from the fact that it simply must hop in frequency, necessitating a frequency-agile oscillator. Unfortunately, since the carrier must span the entire 100 MHz transmit bandwidth of the system at the gigahertz passband, direct-digital synthesis becomes impossible, thus requiring an agile analog synthesizer. The frequency settling time for accurate, low-phase noise phase-locked loops tends to be on the order of microseconds [Stir87], or tens of symbol periods, resulting in a considerable resynchronization delay every time a hop is performed. Likewise, all of the transceivers active within the cell must be synchronized in their hopping scheme, otherwise disastrous cochannel interference will result if two transceivers randomly happen to occupy the same frequency band at the same time.

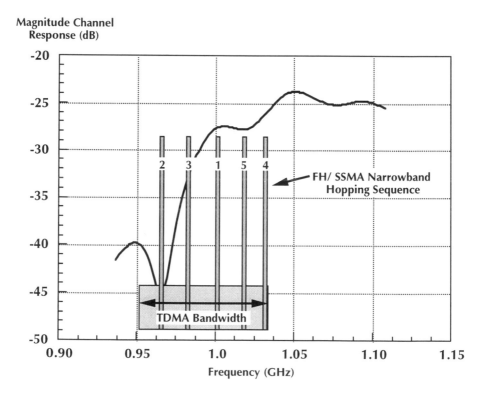

Figure 2.12: Comparison of TDMA and FH/SSMA under multipath conditions

2.3.3. Code-Division Multiple Access

Another multiple-access scheme employs code division, where the users are allowed to transmit simultaneously in the same frequency band, without requiring them to be separated physically by a large distance. Instead, each user is assigned a code waveform from the set $\{p_j(t), j=1..N\}$, where the set represents an orthogonal basis for the signal space and each $\{p_j(t)\}$ occupies B Hz of bandwidth. Since the set $\{p_j(t)\}$ is orthogonal,

$$\int_0^T p_k(t)\, p_j(t)\, dt \;=\; \delta(j, k)$$

where δ is a Kronecker-delta function. The assumption is also made that the autocorrelation function of each of the $p_j(t)$ is also a delta function, i.e.,

$$\int_{-\infty}^{\infty} p_j(\tau) p_j(t + \tau)\, d\tau \;=\; \delta(t)$$

Thus, the transmitted waveform for user m is then given by:

$$2\mathbf{Re}\left\{ \sum_{k=0}^{\infty} A_{kT}\, P_m(t - kT)\, e^{j\omega_c t} \right\}$$

Since these signals are transmitted at the same time in the same frequency band, the received signal at any receiver is just the linear superposition over all N users. Now, after the signals have traversed the channel and have been mixed down to baseband, the receiver is presented with the problem of extracting that particular user's data from the superposition of the data from many users and additive noise, i.e.,

$$\sum_{n=1}^{N} \sum_{k=0}^{\infty} A_{kT}^{n}\, p_n(t - kT) + n(t)$$

However, this is simply accomplished: the receiver knows *a priori* which pulse $p_m(t)$ was used to transmit. Hence, by simply correlating the received signal against the known $p_m(t)$, the correct data A_{kT}^{m} for user m can be extracted, since:

$$\int_{kT}^{(k+1)T} \left\{ \sum_{n=1}^{N} \sum_{k=0}^{\infty} A_{kT}^{n}\, p_n(t - kT) + n(t) \right\} p_m(t - kT)\, dt$$

$$= A_{kT}^{m} + \int_{kT}^{(k+1)T} p_j(t - kT) n(t)\, dt$$

As the noise signal is assumed to be nearly white, its power must be evenly spread out among each of the code waveforms, since these represent a basis for the signal space; hence, the power in the noise term after correlation has been reduced by approximately a factor of N. This is an elegant scheme for achieving multiple access, since it simply performs a correlation in the receiver against a known digital code sequence. In particular, the code space can be designed to consist of binary antipodal sequences of $\{+1,-1\}$ values. Such code sequences are easily synthesized in the transmitter; what is more important is that this implies that the multiplication required in the receiver correlator can be performed easily with a minimum of hardware, even at the speeds required by the downlink.

Essentially, direct-sequence CDMA utilizes pseudo-noise spread-spectrum techniques to achieve its ends, since the original user bandwidth has been expanded to occupy the entire bandwidth available. Alternatively, this can be viewed as multiple high-data rate symbols being sent in place of one user symbol, which effectively is a form of coding redundancy. This increase in error immunity is called the processing gain P_g, defined as the ratio of the spread spectrum to the original spectrum, and is exemplified by the apparent reduction in the noise power after the correlation despreading [Simon85, Cook83, Pick82]. However, beyond simply providing a multiplexing technique, the coding and processing gains of spread-spectrum also yield two other interesting properties: spread-spectrum signals are low-probability-of-intercept, since knowledge of the precise code used is required, and spread-spectrum signals tend to be naturally immune to multipath distortion [Lee88, ProakB83]. Both of these are excellent advantages considering the indoor transmission environment, and the immunity to multipath is especially important, since it implies that a complex adaptive equalizer may not be needed.

The immunity to multipath distortion can be understood by the increased time-resolution of a spread-spectrum signal. Since the $p_j(t)$ spreading function is assumed to have a near delta-function autocorrelation, the arrival time of each symbol can be determined precisely, by using a "sliding correlator" to detect when the correlation is nonzero. For a single transmitted symbol at t=kT, this sliding correlation is given by:

$$\int_{-\infty}^{\infty} A_{kT}\, p(\tau - kT) p(\tau - t)\, d\tau$$

This is clearly nonzero only when t=kT, as expected. However, suppose the signal has been subject to multipath distortion: in particular, it has been passed through a channel whose impulse response is finite in time and given by (from Section 2.1.4):

$$h(t) = \sum_{k=0}^{n} a_k \delta(t - t_k) e^{j q_k}$$

Then the sliding correlation yields a nonzero value not only at t=kT, but also at all of the multipath arrival times $\{t_k\}$ thereafter as well. Hence, the arrival times of the signal can be resolved precisely. In fact, if multiple correlators are used, one correlator per multipath arrival, the data in the multipath itself can be resolved to create a form of time-diversity. Ef-

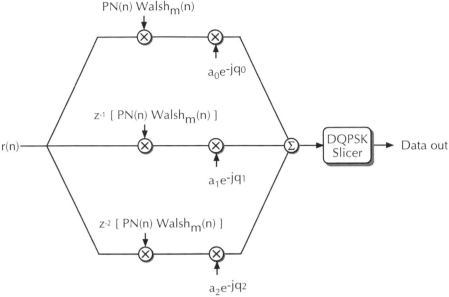

PN(n) Walsh$_m$(n)

$a_0e^{-jq_0}$

z^{-1} [PN(n) Walsh$_m$(n)]

r(n)

$a_1e^{-jq_1}$

z^{-2} [PN(n) Walsh$_m$(n)]

$a_2e^{-jq_2}$

DQPSK Slicer → Data out

$a_0e^{-jq_0}$, $a_1e^{-jq_1}$, $a_2e^{-jq_2}$ are the complex conjugates
of the multipath channel taps.

Figure 2.13: Discrete-time RAKE receiver architecture (3rd order)

The notation PN(n)Walshj(n) indicates a particular family of codes (which will be discussed
in Chapter 3). The subscript m indicates the mth code in the family, corresponding to the
mth user. Since the RAKE combining is usually done digitally, the signals are indicated as
being functions of "n" (discrete) instead of "t" (continuous).

fectively, the multipath reflections are scaled and time-shifted copies of the transmitted
information; by combining these copies in a coherent fashion, the SNR of the system can
be improved. Under white-noise interference assumptions, this becomes the basis for the
RAKE receiver [Dixon84]. A basic RAKE structure is shown in Figure 2.14. For each multipath arrival, a correlator is used to resolve the information available. By multiplying each
estimate against $a_ke^{-jq_k}$ (the complex conjugate of each multipath coefficient) and summing, the estimates are phase coherent with respect to one another, and the receiver SNR
has been increased by a factor equal to:

$$SNR_{increase}(dB) = 10\log\sum_k \left|\frac{a_k}{a_0}\right|^2$$

The number of multipath estimates used is the order of the RAKE, and each path multiplication shown in Figure 2.12 is called a "finger" of the RAKE.

However, the derivation of the RAKE receiver depends upon certain assumptions: namely, the noise in each finger of the RAKE has a fixed, uncorrelated level of background noise. This is certainly true for many of the applications for which spread-spectrum has been applied to date: as an anti-jamming form of military communications. As a multiple access technique, with multiple users present, the assumption of a fixed background white noise level in each finger is violated, since the signal from other users can interfere with the desired signal. In fact, it can be shown that, under mild assumptions, the noise N(i) present in the i-th finger of the RAKE structure has variance given by [Teus97]:

$$Var(N_i) \propto \left(\frac{N}{M}\right) \sum_{k = 0, \, k \neq i}^{L - 1} |a_k|^2$$

where N is the number of users within the cell, M is the processing gain, and L is the order of the RAKE. This implies that the noise power is different at each finger of the RAKE structure, and hence the structure of Figure 2.12 is not optimal. The exact combining strategy must be dependent not only on relative signal strength (as is true in Figure 2.12), but also on the relative noise energy in each finger. The development of an interference-limited RAKE structure is ongoing [Teus97], as well as advanced forms of combining involving simultaneous detection across multiple users [Verdu93, Verdu97].

In either the white-noise limited or interference-limited case, knowledge of the $\{a_k\}$ multipath coefficients is required to achieve an improvement in SNR, since the interference signal must be determined. The specific performance system implementation issues behind determining $\{a_k\}$ will be deferred to the next chapter, where a scheme to continuously monitor the $\{a_k\}$ coefficients will be described. However, some fundamental limitations must be considered beforehand: in particular, the resolvability of the $\{t_k\}$ in the time domain. The $p_j(t)$ signals described above are idealized, continuous-time signals with delta-function cross correlation. In practice, the $\{p_j(t)\}$ are bandlimited, having been generated by digital logic and then filtered; it is impossible to achieve delta-function cross correlation with such signals. In fact, the signals are bandlimited to the smallest fraction possible above $1/T_{chip}$ possible, because of spectral efficiency concerns. Thus, the $\{t_k\}$ can be resolved to no better than T_{chip} in time by the receiver; if several arrivals happen within a single period T_{chip}, they can be considered to blur into a single arrival: these are called the resolvable arrivals. The number of resolvable arrivals is clearly related to the delay spread T_{del}; in particular, it is equal to $(1+T_{del} / T_{chip})$. The greater the spreading factor, the more paths can be resolved and hence the greater immunity to multipath. However, to once again illustrate this resolvability-bandwidth tradeoff, for a fixed user data rate, a larger spreading factor implies greater bandwidth!

From a practical standpoint, the use of spread-spectrum has been traditionally hampered by an intrinsic set of problems: synchronization required in the receiver, efficiency in spreading code design, and near-far crosstalk. Surprisingly, each of these problems can be eliminated by the fact that the broadband downlink signal is being transmitted from a single point of origin: namely, the base station in each cell:

- In the above analysis, it was assumed that the correlation performed in the receiver is perfectly aligned with the incoming signal. If they are not synchronous, the scheme fails since the receiver only sees the partial correlation, which is designed to be small, and thus results in severe information loss. Since a base station is available, a pilot tone can be injected at the transmitter to aid the mobile timing recovery, avoiding many of the difficulties with complex matched-filter based algorithms.

- The design of the codes used, i.e., the set of orthogonal signaling waveforms, is another difficult issue that has received extensive discussion in the literature [Gold67, Sarwa80, Scholtz79]. Perfectly orthogonal codes with perfect auto-correlation are difficult to generate; hence, all practical codes always have some nonzero correlation between the $\{p_j(t)\}$ transmit waveforms. Without the ability to achieve code synchronization between all users, the available set of orthogonal codes is fairly limited; Gold and Kasami sequences [Kasami66, Gold67] are the best known candidates. The coding efficiency with such sequences is quite poor; for a spreading factor of 64, there are only 21 known Kasami sequences available, leading to an efficiency of 33%. However, with code synchronization between all users possible in the base station, the family of Walsh functions [Walsh73] can be employed, leading to an effective code efficiency of 100% (64 codes for a spreading factor of 64).

- Nonzero cross correlation in conventional spread-spectrum systems leads to near-far crosstalk problems, where a transmitter close to the receiver can corrupt the data from one further away despite code orthogonality, since the received power from the nearby transmitter is much higher, and hence the undesired correlation may mask out the actual signal. However, since all signals are originating from a single point, no near-far reception problems can occur.

Lastly, if direct-sequence CDMA is to be employed, small constellations in the QAM modulation are dictated. The use of larger constellations, while reducing symbol rates, significantly decreases the tolerable signal to noise ratio at the receiver. The situation with CDMA is unique; as transmit power scales, so does the noise, since the vast portion of the noise interference comes from other users. Due to nonzero correlations and other nonideal effects, even greater spreading factors and processing gain will be required for large constellations, thus offsetting the decrease in symbol rate. It has been found that QPSK provides a reasonable constellation configuration for use in direct-sequence CDMA [Simon85, ProakB83].

2.4. Summary

With the above discussions of multiplexing and modulation, the various advantages and disadvantages must be considered in light of the characteristics of the indoor picocellular propagation environment, and the data rates required by the personal communications system. The real advantage of CPM techniques, power efficiency in the transmitter, is negated by the fact that the transmit distances involved with a picocellular scheme are small, with very low transmit power requirements. Given the poorer spectral efficiency of CPM, QAM is thus the logical choice for modulation, especially if differential phase encoding is employed to alleviate carrier recovery requirements. Under the same considerations, the equalization difficulties and overhead costs incurred by TDMA do not bode well for its use in a system that already requires user data rates of 2 Mbps. Frequency hopping is an attractive compromise between TDMA and direct-sequence CDMA, possessing multipath immunity while only requiring minimal equalization in the received signal; however, it has severe analog oscillator requirements in the receiver, as well as significant overhead incurred by the need to resynchronize after each hop. The natural immunity of CDMA to multipath, and the intrinsically digital nature of the spreading codes makes it an ideal choice for a broadband wireless indoor communications system; we will see later that the use of CDMA, with its ability to intrinsically reject additive noise, also results in significant reductions in analog hardware complexity.

Thus, with the transmission environment in mind, differential QPSK coupled with code division multiple access is the candidate of choice for the downlink implementation. The exact link specification, based on these decisions, will be discussed in the next chapter.

REFERENCES

[Black91]   K.L. Blackard et al. "Radio Frequency Noise Measurements and Models for Indoor Wireless Communications at 918 MHz, 2.44 GHz, and 4.0 GHz." Submitted to *IEEE Transactions on Antennas and Propagation*, June 1991.

[Buck93]   J. Buck. "The Ptolemy Kernel: A Programmer's Companion for Ptolemy 0.4," UCB/ERL Memorandum No. M93/8, January 19, 1993.

[Cook83]   C.E. Cook and H.S. Marsh. "An Introduction to Spread Spectrum," *IEEE Communications Magazine*, pp. 8-16. March 1983.

[Dixon84]   R.C. Dixon. *Spread Spectrum Systems, 2nd ed.*, New York: J. Wiley and Sons, 1984.

[Ekel87]   B. Ekelund and M. Torkelson. "Waveform Generation for CPM Transmitters." Memorandum of the Department of Applied Electronics, University of Lund, Lund, Sweden, January 1987.

[Feher87]   K. Feher. *Advanced Digital Communications*. New Jersey: Prentice-Hall Inc. 1987.

[Gold67]   R. Gold. "Optimal Binary Sequences for Spread-Spectrum Multiplexing." *IEEE Transactions on Information Theory*, Vol. IT-13, pp. 619-621. October 1967.

[Gold68]   R. Gold. "Maximal Recursive Sequences with 3-Valued Recursive Cross Correlation Functions." *IEEE Transactions on Information Theory*, Vol. IT-14, pp. 154-156. January 1968.

[Kasami66]   T. Kasami. "Weight Distribution Formula for Some Class of Cyclic Codes," Technical Report No. R-285. Coordinated Science Laboratory, Univ. of Illinois, Urbana, IL, April 1966.

[Lee88]   E.A. Lee and D.G. Messerschmitt. *Digital Communication*. New York: Kluwer Academic Publishers, 1988.

[Lee89]   W. C-Y Lee. *Mobile Cellular Telecommunications Systems*. New York: McGraw-Hill Book Co., 1989

[Murota81]   K. Murota and K. Hirade. "Transmission Performance of GMSK Modulation." *Transactions of the IECE of Japan*, Vol. 64-B, p. 1123. October 1981.

[Nichol91]   H. Nicholas III, H. Samueli. "A 150-MHz Direct Digital Frequency Synthesizer in 1.25-um CMOS with -90 dBc Spurious Performance." IEEE Journal of Solid-State Circuits, Vol. 26, pp. 1959-1969. December 1991.

[Ohrvi89]   S.O. Ohrvick et al. "Microcell Radio Channel: Preliminary Report on Indoor Field-Strength Measurements at 900 and 1700 MHz." Memorandum of the Department of Applied Electronics, University of Lund, Lund, Sweden, January 1989.

[Pick82]   R.L. Pickholtz and D.L. Schilling. "Theory of Spread-Spectrum Communications - A Tutorial." *IEEE Transactions on Communications*, Vol. COM-30, No. 5, pp. 855-883. May 1982.

[Proak83]   J.G. Proakis. "Adaptive Equalization for TDMA Digital Mobile Radio." *IEEE Transactions on Vehicular Technology*, Vol. VT-40, No. 2, pp. 323-332. May 1991.

[ProakB83]   J.G. Proakis. *Digital Communications*. New York: McGraw-Hill Book Co., 1983

[Raith91]   K. Raith and J. Uddenfelt. "Capacity of Digital Cellular TDMA Systems." *IEEE Transactions on Vehicular Technology*, Vol. VT-40, No. 2, pp. 323-332. May 1991.

[Saleh87]   A.M. Saleh and R. A. Valenzuela. "A Statistical Model for Indoor Multipath Propagation." *IEEE Journal of Selected Areas in Communications*, Vol. SAC-5, No. 2, pp. 128-137. February 1987.

[Sarwa80]   D.V. Sarwate and M.B. Pursley. "Crosscorrelation Properties of Pseudorandom and
                 Related Sequences." *Proceedings of the IEEE*, Vol. 68, pp. 593-619. May 1980.

[Scholtz79]   R.A. Scholtz. "Optimal CDMA Codes." *1979 National Telecommunications Conf.
                 Record*, Washington, D.C., pp. 54.2.1-54.2.4. November 1979.

[Seidel89]   S.Y. Seidel and T.S. Rappaport. "Radio Channel Models in Manufacturing
                 Environments." *Wireless Information Networks Workshop*, Rutgers University, Rutgers,
                 New Jersey, June 15, 1989.

[Seidel91]   S.Y. Seidel and T.S. Rappaport. "914 MHz Path-Loss Prediction Models for Indoor
                 Wireless Communications in Multi-floored Buildings." Submitted to *IEEE Transactions
                 on Antennas and Propagation*, May 17, 1991.

[Sheng91]   S. Sheng. *Wideband Digital Portable Communications: A System Design.* M.S. Thesis,
                 Memorandum No. UCB/ERL M91/108, U.C. Berkeley, Berkeley, CA,
                 December 3, 1991.

[Simon85]   M.K. Simon, et al. *Spread-Spectrum Communications, Vol. 1-3.* New York:
                 Computer Science Press Inc., 1985.

[Stir87]   R.C. Stirling. *Microwave Frequency Synthesizers.* New Jersey: Prentice-Hall Inc., 1987.

[Sven90]   L. Svennson. *Implementation Aspects of Decision-Feedback Equalizers for Mobile
                 Telephones.* Ph.D. Dissertation, Department of Applied Electronics,
                 University of Lund, Lund, Sweden, May 1990.

[Teus97]   C. Teuscher. Ph.D. thesis on Multi-Access Limited, Multi-User Combining Techniques,
                 Forthcoming. U.C. Berkeley, Berkeley, CA, 1997.

[Verdu93]   S. Verdu. "Multiuser Detection." *Advances in Statistical Signal Processing Vol. 2: Signal
                 Detection*, pp. 369-409. Greenwich, CT: JAI Press, 1993.

[Verdu97]   S. Verdu. "Demodulation in the Presence of Multiuser Interference: Progress and
                 Misconceptions." *Intelligent Methods in Signal Processing and Communications*,
                 D. Docampo, A. Figueiras-Vidal, F. Perez-Gonzalez, eds, pp. 15-44.
                 Boston: Birkhauser, 1997.

3 System Overview: The Broadband CDMA Downlink

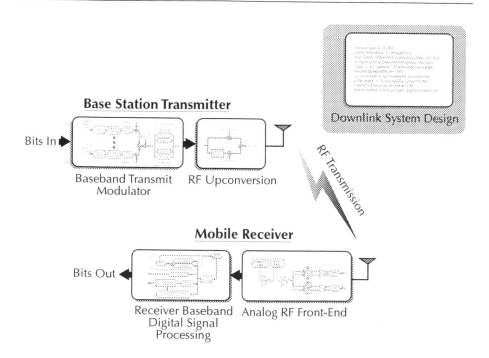

From the preceding discussion, the ability of a synchronous CDMA system to "reject" additive noise and multipath interference, combined with potential reductions in implementation complexity, makes it the multiple access strategy of choice for the broadband downlink. It will become readily apparent that there is a tremendous amount of interdependency between the various parameters in the link design; it is the goal of this chapter not only to describe the parameters themselves, but also to make these interdependencies clear. In Table 3.1, the link specification is given. The user layer parameters - a 2 Mbps per-user data rate and a 2 to 5 meter cell radius - are established by the functionality that we wish to achieve: an indoor mobile picocellular system, capable of providing ubiquitous network access for users with sufficient per-user bandwidth to sustain full-motion digital video. The direct consequences of this – the physical layer and multiple access parameters – are addressed in depth below.

User Parameters	
Per-user data rate	2 Mbps
Reuse pattern	K=7
Reuse type	PN code isolation from cell-to-cell
Cell radius	2-5 meters
Physical Layer Parameters	
Modulation	Differential quadrature phase-shift keying
signaling rate	1 MHz
Carrier frequency	1.088 GHz
Transmit power	0 dBm
99% Power bandwidth	85 MHz
Transmit pulse shape	30% Excess bandwidth raised-cosine
Multiple Access Parameters	
Multiple Access	Direct-sequence, code division multiplex
Processing gain	64 (18 dB)
Aggregate downlink data rate	128 Mbps
Maximum users/base station	64
Spreading Code	64-length Walsh with pseudonoise overlay
Per-user power control	Linear 6-bits (36 dB)
Pilot Tone	32768-length pseudonoise sequence
Synchronization lock time	128 milliseconds, worst-case

Table 3.1: System link parameters

3.1. Physical Layer Parameters

3.1.1. Modulation

As discussed in Chapter 2, the spectral efficiency of QAM is extremely desirable; in light of the need for noise immunity to maximize performance of direct-sequence spread-spectrum systems, a constellation size of 2 bits/symbol (QPSK) will be used. The fact that the

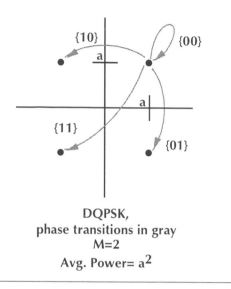

**DQPSK,
phase transitions in gray
M=2
Avg. Power= a^2**

Figure 3.1: DQPSK phase transition diagram

data is entirely encoded on phase, with a single amplitude level, greatly simplifies the re-ceiver design in the sense that extremely accurate gain control is not needed; the auto-matic gain-control loop, comprising of a power detector plus a variable-gain amplifier, needs to be accurate only to a 1-2 dB. Even if only two amplitude levels were permitted, a 1-2 dB amplitude error would be disastrous with regard to system performance, espe-cially with multipath fading taken into account.

Furthermore, to simplify the receiver design further, we remove the need for absolute phase recovery by employing a differential phase encoding (DQPSK); as shown in Figure 3.1, the data is encoded on transitions between symbols, not on the symbols themselves. So long as the channel phase is slowly varying between consecutive symbols, incoherent detection can be accomplished, without needing to know the absolute phase reference. In some sense, the system becomes self-referential; the phase reference for the next symbol is the previous received symbol. If incoherent detection is thus employed, a penalty for this is a 3 dB decrease in signal-to-noise ratio (SNR), since the noise energy from two consecutive symbols is now introduced at the slicer. With an indoor environ-ment, the channel phase rotation will definitely be varying slowly with respect to the sig-naling rate. Channel variation occurs with the movement of people and objects in the environment, on the order of seconds; the signaling rate will be at 1 MHz per user under DQPSK, thus eliminating the need to perform phase recovery.

The interesting phenomenon is that, with a sufficiently high signaling rate, the need for carrier recovery is similarly eliminated. Carrier recovery is needed to resolve frequency mismatch between the transmit local oscillator and the receive oscillator; by examining several families of crystal-referenced oscillators, the error can be discerned. Readily available, off-the-shelf phase-locked loop synthesized oscillators [Phil95, ZComm95] can be toleranced to a worst-case error of approximately 10 parts per million. At 1 GHz, this translates to an error of 10 kHz between transmit and receive oscillators. With narrowband systems, with a per-user signaling bandwidth of 20 kHz, it is obvious why carrier recovery is necessary. However, at a signaling bandwidth of 1 MHz, a 10 kHz error translates to a worst-case symbol-to-symbol differential phase offset of 1%, or about 3°. Compared to the interference energy due to multipath, thermal noise, and inter-cell interference, this is practically negligible. Thus, carrier recovery, which is expensive in terms of hardware and power, can be avoided in the receiver.

3.1.2. Pulse Shaping and System Transmit Bandwidth

Given that wireless systems are intrinsically bandlimited, the decision on what transmit pulse shape to use needs to be determined, and the system spectral usage. From the preliminary analysis in Chapter 2, an aggregate bandwidth of 100 MHz is expected; the exact value will now be determined. From a K=7 reuse pattern with 8 users per cell requiring the full 2 Mbps video data rate (as discussed in Chapter 2), under DQPSK encoding (1 Mbaud/user) the net throughput for all users will be 56 Mbaud. Again, this required throughput will be independent of the multiple access strategy chosen; it is simply a function of the desired per-user throughput and the expected number of simultaneous links per base station. To provide some margin for excess, as well as simplify spreading code generation (Section 4.1) for direct sequence multiple access, an aggregate signaling rate of 64 Mbaud is chosen.

Since spectral efficiency is desirable in such a broadband system, minimizing the baud/Hz ratio becomes paramount, and is a direct function of the excess bandwidth (defined as the fraction above 1 baud/Hz necessary to reliably transmit the signal) of the chosen transmit pulse shape. A simple analog first-order filter will not suffice: to achieve even reasonable intersymbol-interference (ISI) performance, it requires excess bandwidths well above 100%, which is not feasible for this system. Fortunately, complex DSP is available as a synthesis solution for such a pulse; there are no real limitations from implementation complexity, beyond that of available die size. Clearly, the pulse should achieve zero intersymbol interference; it has been shown [Lee88] that such pulses must satisfy the frequency-domain Nyquist zero-ISI criterion:

$$\frac{1}{T_{baud}} \sum_{k} P\left(j\omega - \frac{jk2\pi}{T_{baud}}\right) = 1$$

The family of raised-cosine pulses, given in the frequency domain as:

$$P(j\omega) = \begin{cases} 1, & 0 \leq |\omega| \leq (1-\alpha)\dfrac{\pi}{T_{baud}} \\ 0.5\left(1 - \sin\left(\dfrac{T_{baud}}{2\alpha}\left(|\omega| - \dfrac{\pi}{T_{baud}}\right)\right)\right), & (1-\alpha)\dfrac{\pi}{T_{baud}} \leq |\omega| \leq (1+\alpha)\dfrac{\pi}{T_{baud}} \\ 0, & |\omega| > (1+\alpha)(\pi/T_{baud}) \end{cases}$$

satisfies this, where α is the excess bandwidth and T_{baud} is the baud interval. Indeed, the eye diagram reveals that such pulses do satisfy the zero ISI criterion (Figure 3.2). As an implementation issue, as the excess bandwidth approaches zero, the required number of taps in the filter increases rapidly, to minimize the effects of length truncation in the filter. A reasonable requirement is that the first out-of-band sidelobe due to truncation effects is below 40 dB, similar to goverment regulations for other wireless transmission systems. With this in mind, a plot of FIR (finite impulse response) filter complexity (at a 4 times oversampling ratio over T_{chip}) is shown in Figure 3.3; from this, a choice of 30% excess bandwidth was made, being the largest number of taps that could be implemented using currently available 0.8 micron silicon CMOS fabrication technology.

It should be mentioned that another criterion for selecting the excess bandwidth is given by the timing jitter sensitivity of the pulse shape (the horizontal "opening" of the eye) - timing sensitivity increases with decreasing excess bandwidth, as shown in Figure 3.3. Although the eye width theoretically goes to zero for zero excess bandwidths [Lee88], the probability of this occurring is infinitesimally small given that the pulses have been coded using the combined Walsh-PN sequences. Empirically, it was found that 30% excess bandwidth provided more than sufficient timing jitter immunity: a 1.2 dB amplitude error given the $T_{chip}/8$ worst-case timing error in the receiver (Chapter 8).

3.1.3. Transmit Power and Per-User Power Control

Next, the total allowed per-cell transmit power needs to be considered. Under the assumptions of a code-division system, all cells share the same bandwidth at the same time, utilizing different orthogonal codes to achieve cell-to-cell isolation; the concept of "frequency reuse" becomes one of "code reuse". However, code isolation is not perfect; it is limited by the effectiveness of the spread-spectrum coding, or the processing gain, and thus the transmit power is set by interference from an adjacent cell. To examine this, a link analysis is done with respect to the indoor transmission environment. For a picocellular system with a 4 meter cell radius and a system processing gain of 64 (18 dB), the link budget can be given (in dB) as:

$$P_{adjacent} = P_{transmit} - P_{1dBref} - P_{processing\ gain} - P_{wall\ isolation} - 2.6 \cdot 10 \log (cellradius/1.0)$$

10% Excess Bandwidth Raised-Cosine Eye Diagram

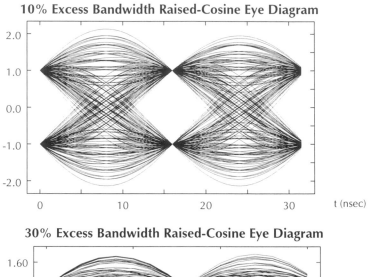

30% Excess Bandwidth Raised-Cosine Eye Diagram

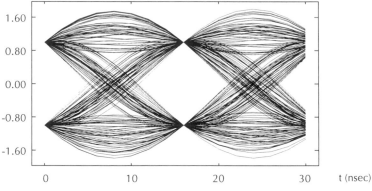

100% Excess Bandwidth Raised-Cosine Eye Diagram

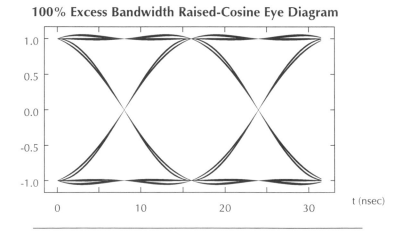

Figure 3.2: Raised-cosine pulse shape eye diagrams.

Note the decreasing sensitivity to timing jitter with increasing excess bandwidth.

P_{1dBref} is the 1 meter path loss reference for the antennae used in the system; for small dipole whip antennas, P_{1dBref} is approximately 30 dB. Given a conservative power loss through a single wall of 7 dB, and using the median path loss coefficient of 2.6, the edge-of-cell interference power from an adjacent cell is thus:

$$P_{adjacent} = P_{transmit} - 30 \text{ dB} - 18 \text{ dB} - 7.0 - 2.6 \cdot 10 \log (4.0/1.0)$$

The processing gain plus wall isolation provide 25 dB of rejection; since the edge-of-cell is being considered, the portion of $P_{adjacent}$ given by $P_{transmit} - 30 \text{ dB} - 2.6 \log(4/1)$ will actually be equal to the nominal (no multipath) in-cell transmit power. Alternatively, this can be interpreted as the interference component from a single adjacent-cell is no more than 0.3% of the total received power (in-cell plus adjacent cell) under free-space conditions. One might question the validity of this, especially (as shown in Chapter 2) the received in-cell power under fading can be significantly below the free-space level. The use of the free-space value can be justified in this fashion: the question here is that of maximum per-cell transmit power $P_{transmit}$, which impacts the system on a global scale. Fading, on the other hand, happens on a per-user, time-varying basis. The system cannot increase the in-cell transmit power simply because one user is in a deep fade; to do so would increase the interference on all of the other users in the system. Thus, the free-space received power is used above from a fairness standpoint; on the average, a user will receive power equal to $P_{transmit} - 30 \text{ dB} - 2.6 \log(4/1)$ at the edge of the cell. As will be discussed below, other system-level elements are employed to combat fading at the user level, such as per-user power control [Yun94] and variable error-correction coding.

In order to optimize capacity, the transmit power within a cell should be chosen such that the adjacent-cell interference power equal to the thermal noise power at the receiver, which as described below is limited by the performance of the input amplifier. In this way, neither the cochannel interference nor the receiver noise is allowed to dominate, yielding a maximum SNR utilizing the minimum transmission power. For a high-performance, low-noise amplifier, a noise Figure as low as 5 dB can be realized, relative to a 50Ω load, resulting in a noise power of -87.5 dBm over a bandwidth of 1 MHz. Using the link power equation above, this implies that $P_{transmit}$ is nominally set at -16.8 dBm, or 0.020 mW; an upper transmit power limit is thus set at 0 dBm, to provide some margin for larger coverage areas. It should be mentioned that this is significantly smaller than the +30 dBm used in conventional analog cellular. As a rather striking demonstration of the impact of system design on hardware, this alleviates the design problems of achieving high-power RF amplification, which has been a critical issue in traditional RF systems.

The above sets the aggregate in-cell transmit power; however, provision must be made to control the power of each individual user's signal. From recent research in spread-spectrum systems, maximization of overall system capacity and performance can be achieved if the amplitude of each user's transmit signal can be controlled as a function of that user's demanded data rate [Yun94, Yun95]. The capacity of a spread-spectrum system can be intuitively thought of as a division in signal-to-noise between users; each user interferes through multipath with other users in the cell. In order to for an individual user to

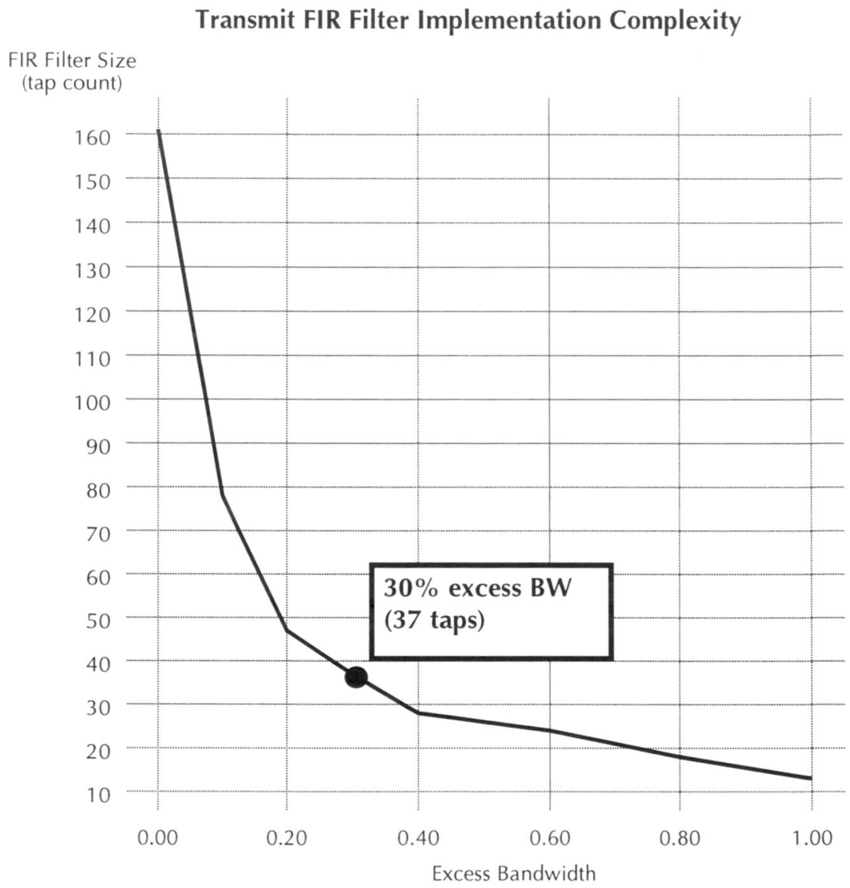

Transmit FIR Filter Implementation Complexity

FIR Filter Size
(tap count)

30% excess BW
(37 taps)

Excess Bandwidth

Figure 3.3: Transmit filter complexity versus excess bandwidth.

achieve higher data rates while preserving error rate, a higher SNR will be needed for that user's signal. Conversely, if a user is demanding lower data rate, that signal can be received successfully with a smaller SNR provided proper error coding is performed. Thus, a user demanding 2 Mbps will cause more interference in the overall system than a user demanding 2 kbps. By taking advantage of this disparity, the overall capacity of the system can be maximized. Likewise, this concept can be extended to a per-application basis: video data may be more immune to bit errors than still image data, implying that a video stream might be transmitted at a lower SNR than a still image.

The idea of varying signal to noise ratio as a function of application, desired data rate, available error correction coding, and overall system capacity is supported in the system specification, with a per-user power control range of 36 dB, linearly variable. The specification is set to meet the needs of ongoing experimentation, based on research that is currently underway [Teus97, Yun95].

3.1.4. Carrier Frequency

Lastly, a decision on carrier frequency needs to be made. Given that spread-spectrum is to be employed, a contiguous block of 85 MHz will need to allocated, in a spectral band that does not already have significant energy. Given that the sheer channel capacity is already in question, clean spectrum will be needed; in-band interferers such as microwave ovens will simply eat into available capacity, disastrously so given the extremely low transmit power in the system. The size of the required spectral band immediately eliminates anything below 1 GHz; however, due to limitations in the available CMOS fabrication technology (a 0.8 micron single-poly process), the carrier frequency cannot significantly exceed this. In Figure 3.4, the spectral energy is shown from 500 MHz to 1250 MHz.

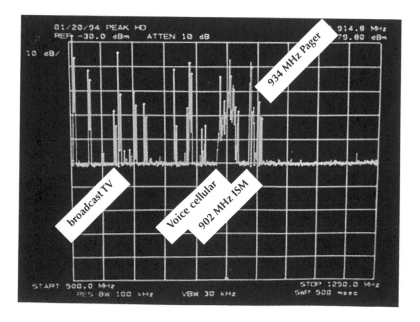

Figure 3.4: Spectral usage, 500-1250 MHz

The significant received energy near 850 MHz is due to analog voiceband cellular; at 934 MHz, a nearby paging transmitter is emitting extremely strong narrowband carrier tones. Between 902-928 MHz, the open spectrum is the so-called ISM (industrial, scientific, medical) band – commercial microwave ovens radiate in this band, as well as digital cordless phones and wireless LAN's. Given that this is unregulated spectrum, the ISM band would appear to be suitable for operation; however, the number of potential local interferers, in addition to the small (26 MHz) size of the band, makes it unfeasible. Above 975 MHz, however, the spectrum is quite clean. Thus, we place the transmit carrier at 1.088 GHz, achievable by the available fabrication technologies. Obviously, this is for prototyping and experimental purposes *only*; government regulatory issues need to be considered if the system is to be deployed in the field. However, for demonstration purposes, the low 1 GHz band is ideal, especially with transmit power levels at or below 0 dBm. The specific value of 1.088 GHz will be made clear in Chapter 6, where the use of sampling demodulation will be discussed and its need for a harmonic relationship between the baseband sampling frequency and the carrier.

3.2. Multiple-Access Parameters

3.2.1. Processing Gain and Walsh Spreading Codes

The key behind any direct-sequence CDMA system is the design of its spreading code; the performance of the code determines how many users can be effectively multiplexed, as well as how robust the system is in the presence of multipath fading. In this system, since the transmit signal comes from a single point-of-origin (the base station), code synchronization is possible between the various users, and this fact will be exploited to its fullest potential. The coding scheme described below – a hybrid PN-Walsh code – is a scaled version of that employed by the IS-95 digital cellular standard [Qual92]; it has been adapted to meet the particular data rate and transmission environment in question. From the system bandwidth analysis, each user bears a signaling rate of 1 Mbaud, with a minimum spread rate of 64 Mchip/sec (64 chips are transmitted per user bit). The required spreading factor may be higher than this, since coding inefficiencies, like those found in the Gold codes, may require a larger factor to accommodate 64 users simultaneously.

Fortunately, a code with perfect efficiency - the Walsh functions - can be employed for this application, and thus the spreading rate for the system is 64 Mchip/sec. The Walsh functions are binary antipodal {+1,-1} sequences that are designed, by definition, to be perfectly orthogonal to one another. In Figure 3.5, several Walsh functions are shown; the Walsh number given for each corresponds to the frequency content, or "sequency", of that function. For example, the zeroth Walsh function is a constant DC value, whereas the 64th Walsh function is alternating +1/-1, being of maximum frequency. Indeed, perfect efficiency is achieved: as shown in Figure 3.6, each user's data is spread by 64, and then all of the spread data streams are linearly combined.

The immediate question arises of why these codes are not used in all direct-sequence CDMA systems. The problem lies in the fact that these codes, while possessing perfect cross-correlation properties, are extremely poor in autocorrelation: they do not appear to be "white" noise. The desirable property of multipath rejection, which depends so heavily on autocorrelation performance, is completely lost. Furthermore, timing recovery in the receiver is exacerbated, since it becomes difficult to determine each user's baseband bit-timing without employing large oversampling factors. Given that a base station is present in the downlink, capable of providing synchronization all of the user Walsh sequences by maintaining a pilot tone, these problems can be overcome.

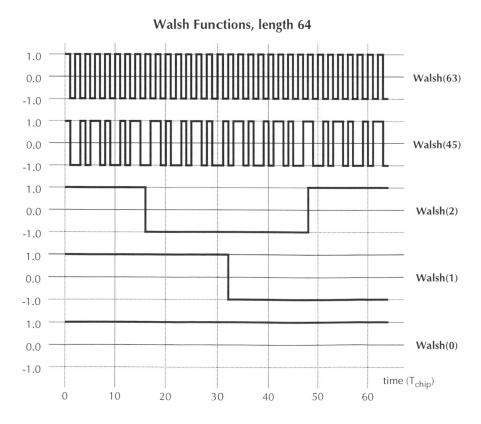

Figure 3.5: Sample Length-64 Walsh Functions

Illustrates the poor autocorrelation properties of the Walsh functions, as well as the need for code-to-code time synchronization (Walsh(2) and Walsh(1) are indistinguishable under time shift.)

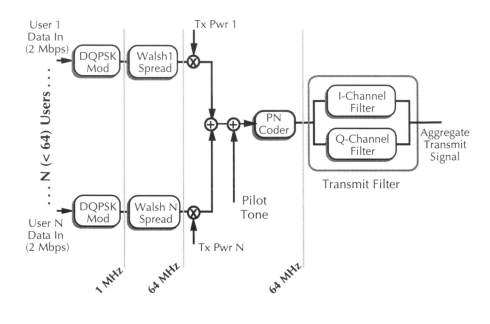

Figure 3.6: Dataflow diagram of the spread-spectrum signal synthesis.
Signal bandwidth at each point is indicated.

3.2.2. PN Overlay Code, the Pilot Tone, and Cell-to-Cell Code Reuse

To restore autocorrelation performance to the overall code, a pseudonoise (PN) code is
multiplied against the spread Walsh sequence on a chip-for-chip basis after all of the user
Walsh sequences are combined, i.e., both the Walsh and PN sequences are time synchronous at 64 MHz (Figure 3.6). The key property of PN sequences is that they appear to be
"white" in nature: their power spectrum is flat, and their autocorrelation is nearly an impulse function (Figure 3.7). Furthermore, the resulting spectrum of the concatenated
Walsh/PN code is also white, to extremely good approximation. Intuitively, it is a similar
situation as the case of sinusoidal modulation of white noise: the periodicity of the Walsh
codes, spectrally concentrated, does not significantly affect the randomness of the spectrally flat PN code. Multiplication in the time domain by a periodic function results in a
shift in frequency, and shifting a white spectrum results in the same. The result is an aggregate code that can simultaneously achieve 100% capacity (with all users are synchronized by the base station), while still maintaining near-white spectral performance.

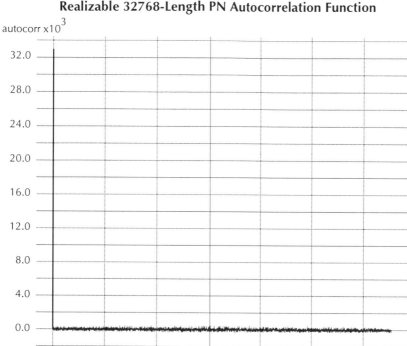

Figure 3.7: Realizable 32768-length PN sequences
(Implementation details of the PN code can be found in Chapter 4)

An important point here is that the length of the PN code and the length of the Walsh sequences need not be the same. The spreading factor is set by the Walsh sequence length: 64. On the other hand, the scrambling code can be significantly longer: as each user's Walsh sequence repeats from bit-to-bit, the scrambling sequence simply continues to progress. The reason that this is important is that longer PN sequences have better auto-correlation properties: the strength of the autocorrelation peak is proportional to the sequence length N. So long as the receiver has some means of synchronizing against the start point of the PN sequence, it can correctly "undo" the scrambling before correlating against the Walsh sequence to recover the user's data.

This synchronization can easily be achieved, given the PN sequence's autocorrelation properties, and this pilot becomes the key to the timing of the entire system. Since the transmit signal in each cell originates from a single point-of-origin, one channel can be expended as a pilot tone, to aid in timing recovery in the receiver. The zeroth Walsh function – a constant - is designated as the pilot, with no data modulation. Examining the pilot-tone signal path in Figure 3.6, the contribution to the output will simply be the constant pilot tone multiplied by the PN sequence, which is effectively just the PN sequence. The receiver can simply search for the strong autocorrelation peak, and thus synchronize its local PN generator with the transmitter's. Since the PN sequence is being transmitted at the spread-spectrum chipping rate of 64 MHz, such synchronization also determines bit-level timing recovery for the receiver; if this pilot tone is continually tracked, any timing deviation can be monitored by chasing the correlation peak, using a delay-locked loop in the receiver. The benefit of a long PN sequence is clear: the effective processing gain on the pilot tone can be significantly larger than 18 dB, since correlation over intervals longer than 64 becomes possible. As the pilot tone is the key to timing recovery in the entire receiver, immunity to multipath and noise is paramount, and the larger effective processing gain greatly reduces the probability of false lock in the receiver.

The length of the PN sequence employed in this system is 32768, or 512 times longer than the Walsh sequence. Beyond the obvious answer of "as long as possible to achieve best code performance", a number of factors contribute to this choice:

- Implementability:
  The length of the PN sequence should be an integer multiple of 64, to give a direct relationship between the PN and Walsh sequences; likewise, it should be a power of two, to employ maximal-length shift-register techniques [Lee88].

- Lock time:
  Ostensibly, search-and-lock times for receiver synchronization are directly impacted by the length of the PN pilot tone An infinitely long PN sequence takes an infinite amount of time to synchronize, as the receiver must step through each PN code phase to determine if lock has been achieved. One might conclude that since lock is only measured at startup, a lock time on the order of seconds could be tolerated. However, if lock is lost, due to channel noise or a severe multipath fade, lock reestablishment time is also critical; a user cannot idly wait for seconds in the midst of a full-motion video stream, or a voice conversation. From the applications envisioned for the broadband downlink, a maximum worst-case lock time of 150 milliseconds was considered tolerable (five video frames at 30 fps).

- Soft Handoff and Cell-to-Cell Code Reuse
  Up to this point, no mention has been made of cell-to-cell code performance, in terms of code reuse. While simple reallocation of Walsh codes is possible, it is far more efficient if the cells are simply designed to be orthogonal to one another. Each base station in the K=7 code reuse pattern uses the same scrambling PN sequence; if a time offset is enforced between the autocorrelation peaks of each base station, the transmit streams from each base station become

naturally orthogonal to one another, given the strong autocorrelation properties of the PN code. Furthermore, this time offset enables the mobile unit to detect nearby base stations: after the mobile synchronizes against the cell's pilot tone, another correlator is kept scanning for autocorrelation peaks. If peaks are detected at time offsets greater than the delay spread of the indoor channel, they must correspond to adjacent base stations. Furthermore, the detected energy in the offset peaks is an indication of how strong, or how close by, the other base stations are. This is the concept of soft handoff [Viter95]; the mobile makes an assessment of the relative received power from adjacent base stations, and thus can assist the system in the decision to make a handoff. In order to achieve this soft handoff, the time offsets between base stations must be guaranteed, by some sort of timing tone on the backbone network. Due to network latency and delays, a synchronization of no better than 2 milliseconds can be achieved, assuming a delay through a single network bridge [Trew90]. At the 64 MHz chipping rate, it implies that a minimum of 4096 chip periods in the PN sequence is necessary to guarantee offset.

From the channel fading statistics, a conservative processing gain of 30 dB is needed for reliable recovery of the pilot tone, implying a code length of at least 1024. However, 1024 does not provide enough time offset slots to accommodate the K=7 reuse pattern given a 2 millisecond synchronization error between base stations; at 64 MHz, a minimum code length of 28000 is needed. From rounding to the nearest power of two, a code length of 32768 is hence dictated.

However, the issue of lock time remains. Assuming a single correlator in the receiver, correlating over the full 32768-long code across 32768 possible code phases, would require a worst-case time of $(32768 \cdot 32768)/64$ MHz = 16 seconds to achieve lock, far too long to be useful. One obvious solution would be to use 107 parallel correlators to achieve the required 150 millisecond lock time; however, this rather sorely violates any reasonable power and implementability constraint in the receiver. Some parallelism is quite possible - as will be discussed in Chapter 8, four parallel correlators are naturally available in the receiver architecture for code search. The key to achieving lock time is the realization that a sufficiently long subsequence of the PN sequence, statistically, should also be nearly white. Instead of correlating over the complete 32768-long sequence, correlating over 1024-long subsequences is sufficient, provided that the spectral characteristics of the subsequences are indeed flat. As will be shown in Chapter 4, the subsequence is nearly white (Figure 4.4); the autocorrelation function exhibits a peak of 1024, as expected, with worst-case autocorrelation error performance 20 dB below the peak. Four parallel correlators, correlating over 1024-long subsequences across possible 32768 code phases, requires a worst-case time-to-lock of 128 milliseconds, which is acceptable.

3.2.3. Channel Estimation and RAKE Reception

The pilot has even greater functionality than simply providing timing recovery for the mobile and mobile-assisted handoff: it can provide the channel estimates necessary to achieve RAKE reception. As described in Chapter 2, direct-sequence spread-spectrum has the ability not only to reject multipath at the receiver, but to coherently combine detected energy from multipath arrivals to increase overall SNR performance, by using some sort of combining structure, either a RAKE structure [Dixon84] or the interference-limited combiner [Teus97]. The key drawback from these techniques is the need to have accurate estimates of the impulse response of the channel, since the relative phasing and amplitude of the arrivals are needed for coherence. In a pilot-tone assisted system, such estimation can be easily performed, without need for adaptive feedback techniques.

The first question that arises is the order, or size, of the combining structure. This is determined by the number of resolvable multipath arrivals, which is controlled by the spreading factor and the delay spread [Linn95], given as:

$$N_{resolvable} = 1 + \frac{T_{del}}{T_{chip}}$$

Intuitively, the time resolution of the spread-spectrum system can be no better than the chipping period T_{chip}; the total number of resolvable multipath arrivals is simply the delay spread over T_{chip}. For the broadband downlink, T_{chip} = 16 nsec, T_{del} = 30 nsec, implying that a third order structure is achievable.

To illustrate multipath resolution, a typical multipath profile for a channel is shown in Figure 3.8, along with the transmitted and received waveform using just the pilot tone (no user Walsh data). As in Chapter 2, this signal can be thought of as the linear combination of time-shifted and attenuated copies of the transmitted pilot tone $s(t)$:

$$r(t) = \sum_k a_{kT}s(t - kT)$$

where the $\{a_{nT}, n = 0..2\}$ are the resolvable multipath coefficients at delay $t = nT$ relative to the first arrival. To resolve the nth multipath coefficient, a 1024-length correlation is performed using a copy of the pilot tone shifted by nT:

$$MPath(nT) = \frac{1}{1024} \sum_{\tau = t - 1024T}^{t} s(\tau - nT) \sum_k a_{kT}s(\tau - kT)$$

$$= a_{nT} + \frac{1}{1024} \sum_k a_k err((n - k)T)$$

where err(t) is the autocorrelation sidelobe noise at offset t (Figure 4.4). To emphasize, this noise is due to the fact that we are taking correlations over a 1024-long subsequence of the 32768 PN code. Figure 3.9a also shows the resulting estimated output; the first and second estimates are quite good, whereas the third estimate is quite poor. This is to be expected; a_{2T} is quite small compared to the first two arrivals, and the estimate is being overwhelmed by intrinsic noise in the code. The 20 dB sidelobe performance of the PN subsequence implies that arrivals that are less than 0.1 in amplitude of the first arrival cannot be resolved. This performance is acceptable; from previous research [Teus94], multipath arrivals weaker than 10 dB relative to the primary arrival are insignificant in both the RAKE and interference-limited combining algorithms. If improving the noise performance of the estimate is required, a simple moving-average filter is quite effective in minimizing the sidelobe noise, given that it is effectively white in nature. In Figure 3.9b, a 4-tap moving-average (MA) filter has been applied to the multipath estimates; SNR has been globally improved by 6 to 10 dB, even estimating the last multipath arrival. It should be mentioned the conditions that the simulation was executed: no intercell interference was present, and a fixed frequency offset between the transmit and receive carriers was assumed (no carrier recovery is performed in the hardware, for power and complexity considerations, thus leading to a significant worst-case carrier offset). The issue of carrier offset compensation is deferred to the discussion on the receiver DSP in Chapter 8; for the moment, it is sufficient that a relative-phase estimation method using subcorrelations has been developed for the determining the channel multipath response. The DSP output is the non-MA filtered estimate of Figure 3.9.

To provide an example of the performance benefit gained by employing a combiner structure, Figure 3.10 shows the BER versus number of users plot for systems with and without a RAKE combiner (a RAKE was chosen for simplicity). Using the Berkeley Ptolemy simulator, the two channels from Figure 2.5 have been simulated for BER as a function of the number of users in the channel, with and without a RAKE combiner. Clearly, as the number of users increases, the multipath interference becomes more and more significant, as an individual user's energy becomes smaller proportional to the multipath reflections. Furthermore, in the case of a severe in-band null, the BER even for small numbers of users becomes unusable, without employing a RAKE combiner. The center curve is the case of the deep-fade channel with a RAKE combiner – for moderate numbers of users in the cell (less than 20), the performance has been dramatically increased, with a decrease in BER by over 4 orders of magnitude. As the number of users increases beyond 20, the RAKE becomes less and less effective, owing to increased noise in the channel estimate itself and loss of coherency. Under these conditions, the interference-limited combiner [Teus97] is more effective; however, Figure 3.10 illustrates the point: a combining architecture will be critical in achieving channel capacity and robust system performance.

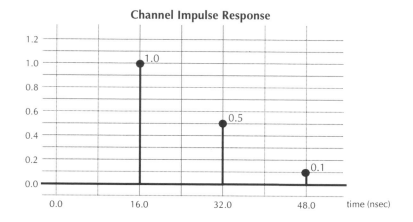

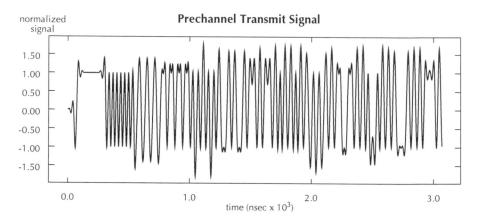

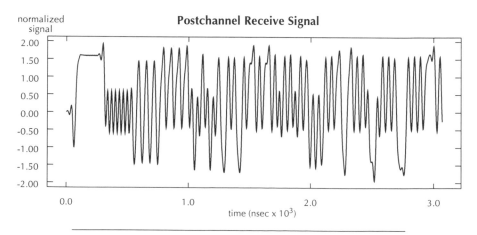

Figure 3.8: Sample multipath channel for channel estimation.
ISI is fairly severe in this case.

Multipath Profile Estimate

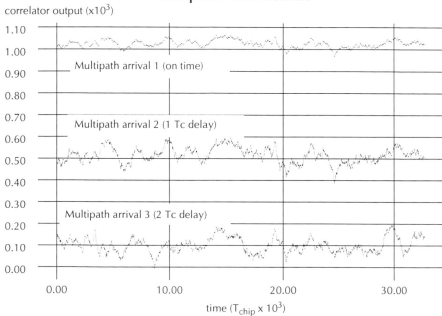

correlator output (x10³)

Multipath arrival 1 (on time)

Multipath arrival 2 (1 Tc delay)

Multipath arrival 3 (2 Tc delay)

time (T$_{chip}$ x 10³)

Figure 3.9a: Multipath estimator output, direct output.

1024-length subcorrelations, 32768-length sequence, Nominal outputs are 1024, 512, 102.4
Comparing to Figure 3.9b, moving-average filtering yields ~6-10 dB SNR improvement.

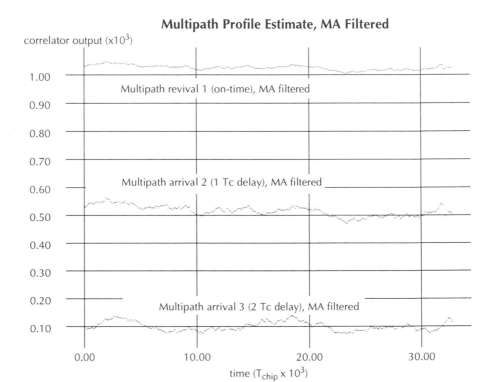

Figure 3.9b: Multipath estimator output, moving-average filtered

1024-length subcorrelations, 32768-length sequence, Nominal outputs are 1024, 512, 102.4
The addition of moving-average filtering yields ~6-10 dB SNR improvement.

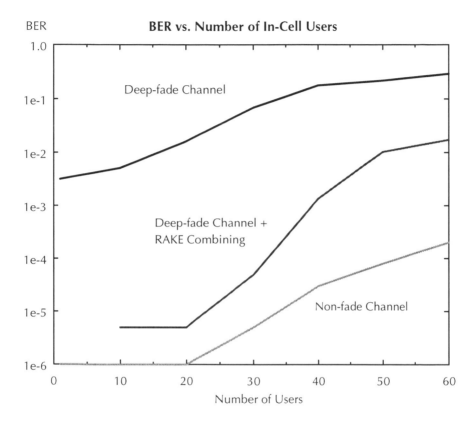

Figure 3.10: Plot of simulated BER versus number of in-cell users, no intercell interference, for the channel profiles shown in Figure 2.4, including RAKE combining. It is clear that with no combining, under worst-case fading, self-interference severely limits the achievable BER performance for even 1 user.

REFERENCES

[Dixon84]   R.C. Dixon. *Spread Spectrum Systems,2nd ed.*, New York: J. Wiley and Sons, 1984.

[Lee88]   E.A. Lee and D.G. Messerschmitt. *Digital Communication.* New York: Kluwer Academic Publishers, 1988.

[Linn95]   J.P. Linnartz. "Wireless Communication Networks: An Intensive Short Course." U.C. Berkeley, Berkeley, Berkeley, CA, 1995.

[Phil95]   W.S. Djen, Philips Semiconductors. *Using the SA7025A and SA8025A for Narrowband Systems, Application Note AN-1890.* Eindhoven, The Netherlands, Jun 18, 1995.

[Qual92]   *EIA/TIA IS-95 Interim Standard, Wideband Spread-Spectrum Digital Cellular Dual-Mode Mobile Station-Base station Compatibility Standard.* Telecommunications Industry Association, April 1992

[Teus94]   C. Teuscher. *Software Simulation for the Infopad Downlink.* M.S. Thesis, U.C. Berkeley, Berkeley, CA, December 1994.

[Teus97]   C. Teuscher. Ph.D. thesis on Multi-Access Limited, Multi-User Combining Techniques, Forthcoming. U.C. Berkeley, Berkeley, CA, 1997.

[Trew90]   G. Trewitt. *Local-Area Internetworks: Measurements and Analysis.* NSL Research Report RR-1, Digital Equipment Corp. Network Systems Laboratory, Palo Alto, CA. March 1990.

[Viter95]   A.J. Viterbi. *CDMA: Principles of Spread Spectrum Communications.* Reading, MA: Addison-Wesley Pub. Co., 1995.

[Yun94]   L. Yun, D. Messerschmitt. "Power Control for Variable QOS on a CDMA Channel." *Proceedings of IEEE MILCOM*, Fort Monmouth, NJ, pp. 178-182. October 2-5, 1994.

[Yun95]   L. Yun, D. Messerschmitt. "Variable QOS in CDMA Systems." *Proceedings of the IEEE ICC*, Seattle, WA. June 18-22, 1995.

[ZComm95]   Z-Communications, Inc. *Application Note AN-106: Using Z-COMM Sub-Miniature VCO's for WLAN Systems.* San Diego, CA. 1995.

4 Transmit Architecture and The Baseband Modulator Chip

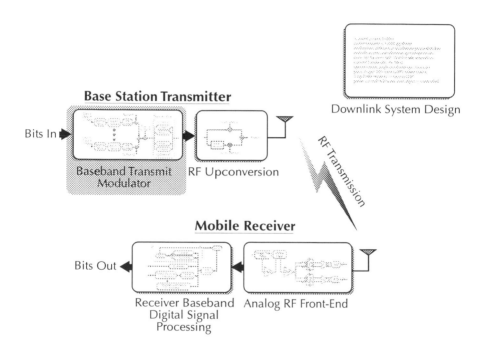

Base Station Transmitter

Bits In ▶

Baseband Transmit Modulator

RF Upconversion

Downlink System Design

RF Transmission

Mobile Receiver

Bits Out ◀

Receiver Baseband Digital Signal Processing

Analog RF Front-End

To begin the discussion of the system implementation, the base station transmit architecture will be described in this chapter and the next. The key functional blocks that need to be implemented are the spread-spectrum digital signal processing (data spreading, scrambling, etc.), baseband pulse shaping, digital-to-analog conversion, filtering, and frequency upconversion to the 1.088 GHz carrier frequency. The hardware partitioning consists of a custom baseband modulator integrated circuit [Peroul94], along with a semicustom board to implement the analog baseband and RF upconversion circuitry [Yee96]. A board-level solution suffices for the transmitter – since this is in the base station, size and power are not nearly as critical, nor is noise performance as extreme an issue.

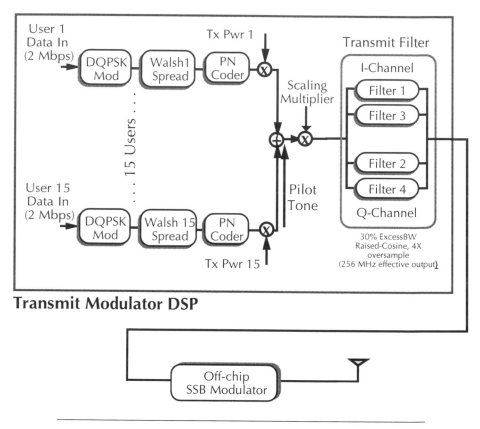

Figure 4.1: Base station transmitter architecture

In Figure 4.1, the transmit architecture for the base station is shown. It implements the multiple-access strategy at its most basic level: Walsh code modulation of each user's data, pseudonoise (PN) sequence scrambling of the overall code-division signal, and pulse shaping. Not surprisingly, the baseband modulator circuit mimics the dataflow block diagram for the overall system (Figure 3.7), with a few subtle differences. The PN modulator, which provides autocorrelation and spectral whitening, has been shifted before the user-combining adder. The pulse-shaping filter, which controls the transmit bandwidth, is implemented as four parallel subfilters instead of a single high-speed one. Choice of number representation is also crucial; the representation changes as the data flows through the chip, from bit input to pulse-shaped output. Each modulator IC only supports 15 users, which is the most common deployment case and minimizes die size; multiple modulators can be ganged in parallel to support more users per base station. A two-stage frequency upconversion is used, one in the digital domain, and one in the analog domain. The impact on hardware complexity of these decisions and others will become apparent throughout; each block will be discussed, in terms of architecture and implementation.

4.1. The Baseband Transmit Modulator

4.1.1. DQPSK encoder

As the first block in the modulator, the differential quadrature phase-shift key (DQPSK) coder implements the bit-to-baud encoding shown in Figure 3.1. Number representation here is extremely important: the possible values for either the in-phase or quadrature symbols are bi-state (+/- 1), and it is natural to map these two values into a single digital "0" or "1". The representation chosen is:

+1 maps into digital "0"
-1 maps into digital "1"

It is akin to a "1-bit" signed magnitude. The ramifications of this will become apparent in the PN and Walsh modulator blocks: multiplies under 1-bit arithmetic can be isomorphically translated into a digital XOR (e.g., -1 • -1 = 1, or "1" XOR "1" = "0"). Under this representation, the DQPSK coder implementation is shown in Figure 4.2.

4.1.2. Walsh Modulator

The encoded symbols are then spread, from the 1 Mbaud rate up to the full 64 MHz spread-spectrum chipping rate. Since 64 individual Walsh codes need to be generated, a 6b-wide datapath is dictated; the architecture chosen is from [Beau84], and is based on Gray-code generators (Figure 4.3). Each of the 64 bits in the Walsh sequence, which is clocked at 64 MHz, is XOR'ed with the complex-valued user data symbol, which is clocked at 1 MHz, to effect spectrum spreading from 1 MHz to 64 MHz.

4.1.3. PN Modulator

After Walsh spreading, the pseudonoise (PN) sequence is then applied for spectral whitening. As in the case of the Walsh modulator, the PN sequence is XOR'ed bit-for-bit against the incoming complex-valued spread data, at 64 MHz. The key question in this block is the actual implementation of the generator – in particular, the synthesis of a 32768-long pseudonoise code. From the discussion on optimal PN sequence length in Chapter 2, a minimum length of 32768 is needed for sufficient cell-to-cell code isolation, as well as being a multiple of the spreading factor for timing recovery.

One well-known class of PN generators is the so-called group of maximal-length shift register (MLSR) sequences. They synthesize PN sequences of length (2^n-1), and are easily implemented using n-bit shift registers in nonlinear feedback (XOR's in the feedback path) [Lee88]. Ergo, a 15-bit MLSR sequence would seem to suffice, generating a 32767-long sequence, and a random bit added at some point. The additional bit will impact the auto-

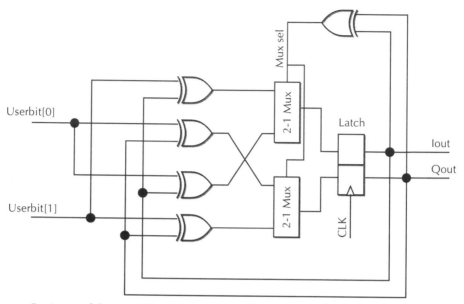

For inputs: If the input bit sequence is: 00011011,
then the inputs are presented as two-bit words: {00},{01},{10},{11}
leftmost bit is Userbit[1].

Figure 4.2: DQPSK coder schematic

correlation performance of the sequence to some extent; however, it should be "nearly" white. Alternatively, a 32768-long subsequence of a 16-bit MLSR sequence would also seem to suffice; again, such a subsequence should be "nearly" white. Quantifying the term "nearly", in terms of impact on system performance, becomes a crucial issue.

The implementations of these two possible choices are shown in Figure 4.4; for comparison, the implementation of a simple 15-bit MLSR sequence generator is also shown. One of the key properties of MLSR generators is exploited here. Since the shift register is n-bits wide, there are 2^n possible words; the MLSR feedback guarantees that each word will occur in the sequence once and only once, with the exception of the all-zero word. By searching for a particular word, an extra bit can be consistently injected into the pseudorandom sequence. In particular, the implementation keys off the "01111....111" word; this will occur one clock cycle before the all-one's word (the extra delay is needed for control circuitry). In Figure 4.4c, an extra clock cycle is added after the all-one's word to inject the extra bit; in Figure 4.4b, the 32768-long subsequence is forced to terminate, by parallel loading a seed into the register. The seed is chosen such that, 32768 cycles later, the all-one's word will occur once again.

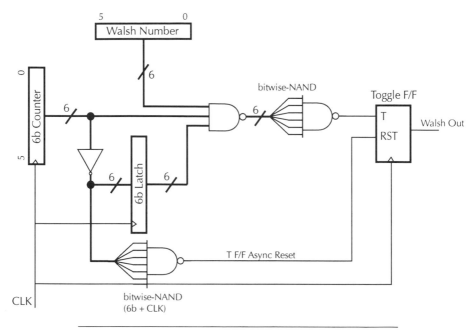

Figure 4.3: Walsh sequence generator (length 64).

MSB-LSB are indicated on all applicable blocks.

In Figure 4.5, the autocorrelation function of each of the generated sequences is shown. For comparison, the autocorrelation function of the true 32767-long MLSR sequence is shown. The term "nearly" is now clear; compared to the peak autocorrelation value, the largest off-peak correlation is less than 512, out of 32768. Given that this is 36 dB down from the peak, from a signal-to-noise standpoint utilizing either the subsequence approach or the additional bit approach has negligible impact. Figure 4.4b was implemented; it has the property that the clock to the PN generator does not need to be gated, eliminating problems with clock skew and glitching.

Lastly, the performance of the code under 1024-long subcorrelations needs to be examined. The need for 1024-long correlations is to minimize the amount of time required to perform functions such as timing acquisition and tracking on the pilot tone, as first mentioned in Chapter 3. By "subcorrelation," it is meant that partial sums of the full correlation are taken; since subsequences are used, two parameters are necessary to specify the subcorrelation, i.e.:

$$\mathrm{subcorr}(t, \tau) \;=\; \sum_{k\,=\,0}^{1023} \mathrm{pn}((\tau - k)\bmod 1024)\,\mathrm{pn}((t - k)\bmod 1024)$$

Basic 15-tap shift register PN sequence generator

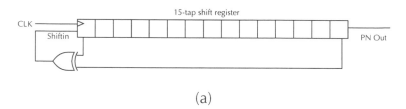

(a)

16-tap shift register base, 2^{15}-length subsequence

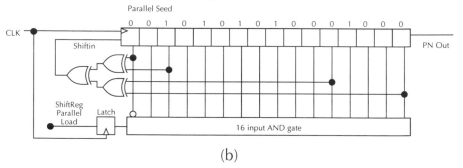

(b)

15-tap shift register base, additional bit added

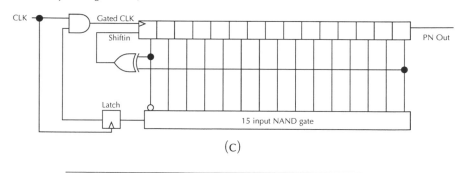

(c)

Figure 4.4: PN sequence generators

(a) True 15-tap, length 32767 MLSR schematic
(b) 16-tap, length 32768 subsequence schematic
(c) 15-tap, length 32768 schematic
Figure 4.4b was implemented.

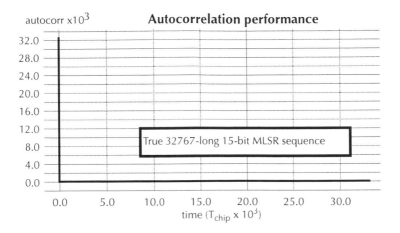

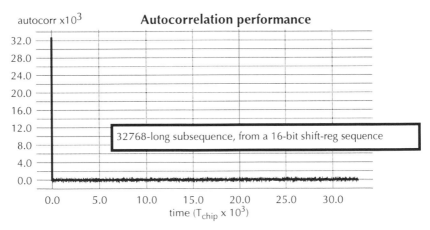

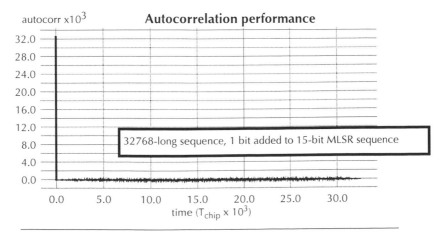

Figure 4.5: Autocorrelation performance curves of the architectures
shown in Figure 4.4.

Autocorrelation performance, using 1024-long subcorrelations

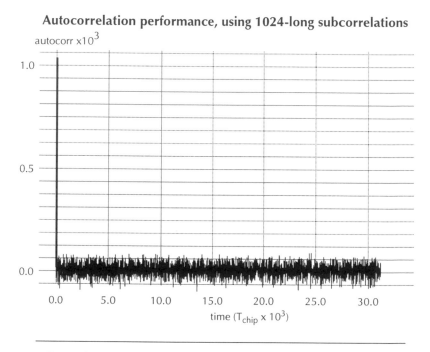

Figure 4.6: Autocorrelation performance under restriction to length 1024 subcorrelations (taken for τ=0; the curve is similar for τ nonzero).

The properties of this 1024-long subcorrelation must be examined carefully, since a subsequence of a pseudorandom sequence is not necessarily itself pseudorandom: for example, with a 15-bit shift register, a subsequence of 15 "1"'s in a row must occur at some point in the sequence, which is hardly random in nature. In Figure 4.6, the plot of 1024-long subcorrelations of the full 32768-long sequence is shown for τ=0; as expected, when the subsequences are synchronized (t=τ=0), a peak value of 1024 is seen. However, the off-peak performance is crucial; when the receiver is searching for the pilot tone to indicate lock, tracking the peak for timing recovery, or measuring the multipath profile, it is this off-peak performance that determines the performance of these functions. From this, the signal-to-noise ratio (SNR) is approximately 20 dB, translating to 3 bits of dynamic range. The choice of 1024 was driven by the need to minimize search-and-lock times, plus the need to minimize complexity in the receiver; under simulation with the multipath channel, 20 dB is the smallest value that reliably achieves lock and timing recovery.

If greater SNR is needed, Figure 4.7 displays the plot of SNR as a function of subcorrelation length – surprisingly, the SNR drops quite slowly, 6 dB for every doubling in the subcorrelation length. This translates to a doubling in the peak amplitude, which is expected; however, the off-peak noise remains fairly constant in amplitude past a length of 256. Hence, for every 6 dB of SNR performance in the autocorrelation code, a doubling of the receiver lock/acquisition time is necessary, along with increasing the dynamic range by 1 bit. As described previously in Chapter 3 with regard to improving channel multipath estimation, the moving average filter technique is applied here to effectively increase the subcorrelation length by a factor equal to the number of taps.

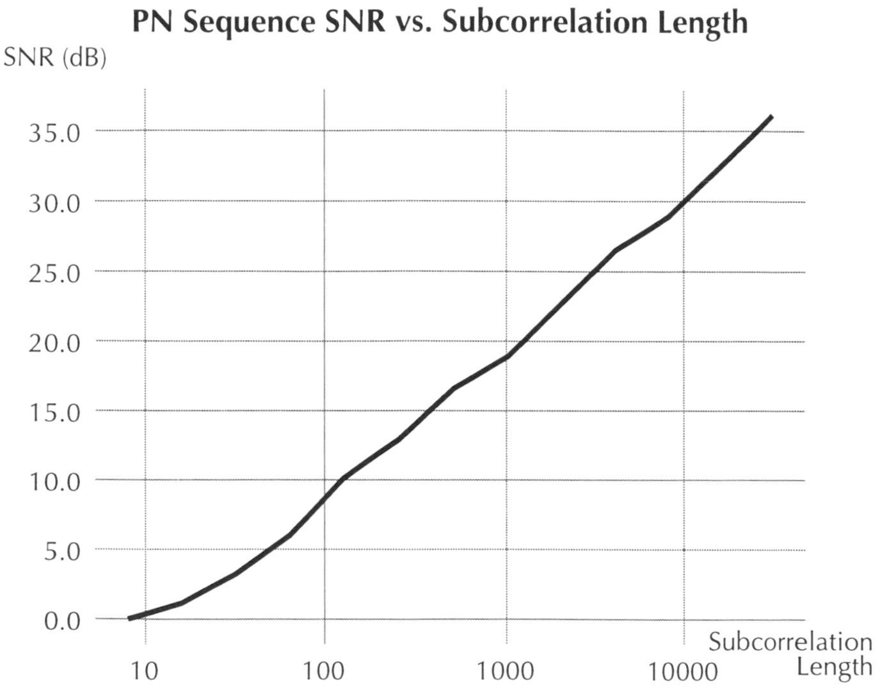

Figure 4.7: Plot of SNR due to off-peak nonzero autocorrelation, as a function of subcorrelation length.

4.1.4. Per-User Power Control

As described in Chapter 3, each user's data stream can be modulated by an power control gain factor to maximize overall system capacity. Given that per-user power control is a topic of current research [Yun95], a dynamic range of 36 dB is set to provide margin for experimentation. This dynamic range corresponds to a 6-bit control word for each user, which modulates the signal amplitude linearly. Number representation here again becomes important. Given that the subsequent stage is the combiner, where all of the user transmit signals are linearly added, a two's complement number representation is desirable. Although the 1-bit signed magnitude representation has served well up to this point, the stream now needs to be translated into an equivalent 6-bit two's complement representation. Instead of implementing a true multiplier, a simple multiplexer can be used: both the power control value and its negative are presented to the input of the muxes, and the incoming 1-bit data on the I/Q lines selects the positive or negative value.

From this, it is clear why the PN modulation was moved before the combining and per-user power control: had it been left post-combiner, as in Figure 3.7, the +/-1 multiply from the PN code would have required the two's complement negation of a 10-bit word at 64 MHz. By simply moving the PN modulator before the per-user power control, a simple XOR of the sign bit suffices.

4.1.5. The Pulse Shaping Filter

At this point, all 15 user data streams and the 6-bit pilot tone are linearly summed; the resulting 10-bit wide datastream is the baseband representation of the transmit signal. The remaining signal processing in the transmitter consists entirely of bandlimiting this signal, digital-to-analog conversion, and frequency translation into the 1 GHz transmit band. As described in Section 3.1.2, a baseband digital bandlimiting filter is desirable to implement a 30%-excess bandwidth raised cosine pulse shape, for zero intersymbol inteference (ISI) at the expense of as little excess bandwidth as possible. Since the incoming data is complex-valued (in-phase and quadrature), two real-valued filters are needed to implement the equivalent complex-valued filter. Given the rather specialized shape of the desired transmit pulse, an IIR (infinite impulse response) implementation becomes problematical; instead, a straightforward transversal FIR (finite impulse response) topology is used. As the impulse response will necessarily be time-truncated, a minimum out-of-band rejection of 40 dB is placed on the filter, which is typical of emission-masking constraints and sets the 99% power containment bandwidth of the transmit signal. The out-of-band reject specification determines the number of taps needed in the filter structure, as well as truncation performance of the taps themselves.

In order to perform pulse-shaping on the 64 MHz stream, oversampling is needed to effectively interpolate the shape between consecutive data points. With the already-high data rate, as small of an oversampling factor is needed; in this case, 4x, with an effective filter output rate of 256 MHz. Under this oversampling constraint, from simulation the total number of taps needed in the filter is 37 (Figure 3.4); after Booth-encoding (canonical signed digit representation) of the tap coefficients and truncation, the resulting filter transfer function is shown (Figure 4.8).

One critical issue here is the dynamic range in the filter; given that digital-to-analog conversion needs to be done at the 256 MHz rate, a high-order digital-to-analog converter (DAC) is out of the question; the highest-performance commercially available bipolar DAC's can provide 10 bits of resolution at these speeds, and no more. Given that the input already has 10 bits of total dynamic range (15 users plus the pilot tone, with 6 bits of power control each), there is an obvious problem here: if all users are present, the full res-olution of power control is not available, given truncation noise in the filter. Clearly, the worst-case situation occurs with 14 users and the pilot set to maximum power, and one user at the minimum power control value; the effective pulse shape of the single user will be nearly at the noise floor of the filter itself. In Figure 4.9, this situation is illustrated graphically: since the data streams are orthogonal and uncorrelated, the total signal power is simply the aggregate sum of the individual signal powers. If the filter were to have suf-ficient truncation noise performance to maintain the 40 dB out-of-band pulse shape for the single user, 100 dB of dynamic range would be required, or over 16 bits. From simu-lation, if some performance loss were allowed in terms of truncation noise injection, the single user could be recovered if the quantization noise floor were limited to 72 dB, or approximately 12 bits of dynamic range.

The ramifications of this, coupled with the 10 bit limitation in the DAC, imply that the least-significant three bits of the input signal to the filter have an amplitude less than the filter noise floor and are ineffective; if a user is set up such that its signal power is below -42 dB relative to the total signal power, that user will be subject to excessive BER at the re-ceiver. To control this dynamic range issue, a "multiplier" is inserted before the filter as a form of aggregate transmit power control. The multiplier is implemented as a 4-bit

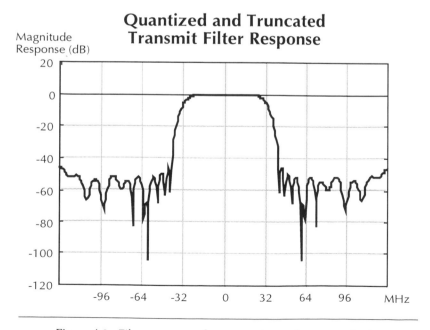

Figure 4.8: Filter response after truncation and tap quantization

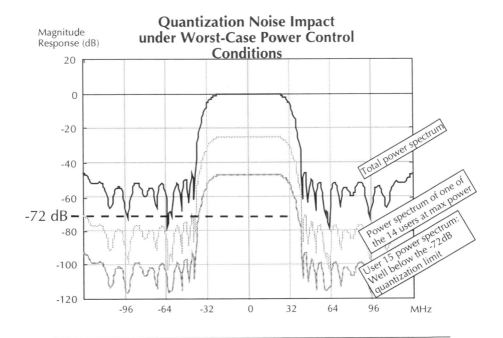

Figure 4.9: Filter quantization noise effects at maximum power control dynamic range (14 users with maximum power, one user with minimum power). The -72 dB mark indicates the minimum quantization performance required to recover user 15.

shifter; although it is only provides power control in fairly crude 6 dB steps, it achieves its desired function. The multiplier shift control and error condition check in the base station should be set according to the following equations:

$$\text{Number of Bits to Shift} = \text{floor}\left(10 - \log_2\left(\sum_i P_i\right)\right)$$

$$\text{Error} = \bigcap_i \left\{ \left(10\log_{10}\left(\frac{P_i}{\sum_i P_i}\right) \right) \le -72\text{dB} \right\}$$

where the P_i's are the users' power control settings (as 6-bit integers) and P_0 is the power in the pilot tone.

4.1.6. Filter Implementation

At this point, the physical implementation of the 256 MHz FIR pulse shaping filter needs to be discussed. Given the technology available (a 0.8 micron standard digital process), achieving these speeds can only be done with careful consideration of the architecture to be employed [Wong91]. A basic transversal structure, as shown in Figure 4.10a, simply cannot achieve a 4 nsec operational cycle time, especially given that all multiplies and accumulates are done with data values of at least 10 bits. Parallelism and pipelining of the datapath need to be included.

4.1.6.1. Upsampling

The immediate optimization that occurs is consideration of the upsampling nature of the filter. The data samples are incident on the filter at 64 MHz; zero-interpolation is the natural means of upsampling the signal. The data stream should be thought of as a train of impulses being fed into a filter with the desired impulse response; the zero-interpolation simply describes the time in between each successive impulse. Thus, three of every four delay registers in the transversal filter structure of Figure 4.10 will have zero samples at any given sample point, and their contribution to the output of the filter need not be computed. The implication is that the structure can be parallelized into four subfilters, running at 64 MHz, and their outputs interleaved with a multiplexer back to the full 256 MHz rate. It is important to note that the data is brought off-chip at 64 MHz, with the interleaving multiplexer implemented using off-the-shelf ECL components. There are two reasons for this: first, driving the output pad ring of the chip at 256 MHz is difficult without incurring tremendous substrate injection and ringing, resulting in questionable signal integrity; second, bringing off the outputs in parallel allows more flexibility in the upconversion architecture, as described in Chapter 5.

4.1.6.2. Tap Symmetry

From the coefficient assignment plot of Figure 4.10, several more optimizations become readily apparent from the structure of the taps themselves. For example, the Nyquist criterion reduces filter 1 into a simple delay; all of the other tap coefficients are zero, reflective of the zero-ISI nature. Furthermore, the pulse shape is symmetric, implying that filter 3 is perfectly symmetric, and filters 2 and 4 are the time-reversed images of one another. An efficient implementation of symmetric filters can easily be achieved. As an example, in Figure 4.11, a 6-tap symmetric filter is shown, along with two architectural minimizations. In the structure, there are 6 taps but only 3 different coefficients: A1, A2, and A3. Originally, when calculating the current output it implements the structure of Figure 4.11a - a simple transversal filter.

By grouping the multiplications with the same tap weights, a much simpler implementation is produced. At the expense of adding an adder in front of every multiplier, the structure can be implemented using half as many taps (Figure 4.11b).

Lastly, the addition at the output of the filter needs to be pipelined, since it impossible to combine add all of the multiplier outputs in a single cycle. The final structure (which was implemented), is shown in Figure 4.11c. The pipelining stages simplify the structure because the they also force an extra delay for the input data in each tap. In the Figure 4.11b, x(n-5), x(n-4), and x(n-3) are fed back; from a layout and routing standpoint, this presents several difficulties, especially given the wide datapaths in question. In the final pipelined structure, only a single signal has to be routed back.

This structure can be verified by comparing figures 4.11b and 4.11c. In Figure 4.11b, the input to the first multiplier is the sum of x(n) and x(n-5); i.e., it must add the current sample along with the sample that appeared five time periods ago. The second tap must add x(n-1) and x(n-4), combining samples with a delay of 3 between them. It can be seen that Figure 4.11c is equivalent; the first tap adds x(n) and x(n-5), maintaining a time delta of 5 samples. Similarly, the second tap combines x(n-2) and x(n-5), maintaining a time delay of 3 samples, with the pipelining in the output adder retiming the entire structure.

Lastly, it should be mentioned that for filters 2 and 4, no similar optimization can be made, as the tap coefficients do not exhibit similar symmetry, beyond the fact that they are time-reflected versions of one another. However, these filter structures must still be pipelined, especially with regard to the output addition, as shown in Figure 4.12.

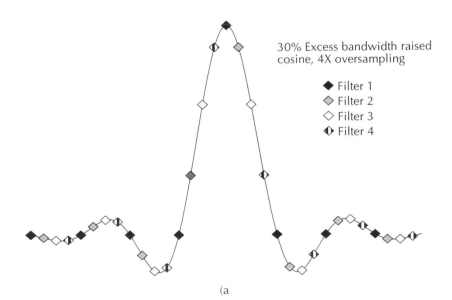

30% Excess bandwidth raised cosine, 4X oversampling

◆ Filter 1
◈ Filter 2
◇ Filter 3
◐ Filter 4

(a

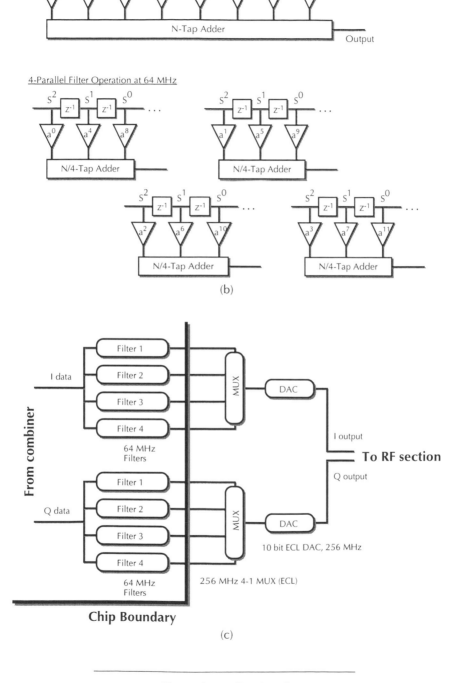

Figure 4.11: Continued.

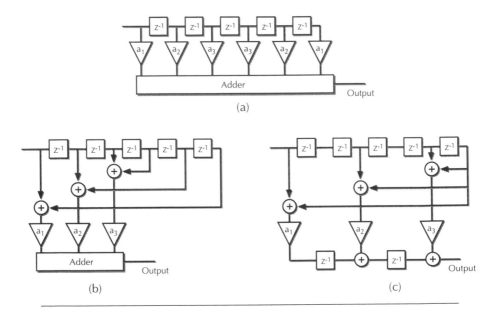

(a)

(b)   (c)

Figure 4.12: Symmetric Filter Minimization (filter 3)

(a) Original transversal structure (b) Shared multipliers (c) Pipelined output adder

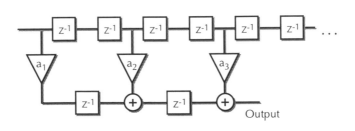

Figure 4.13: Pipelined output adder filter structure

4.1.6.3. $f_s/4$ Modulation

The final filter minimization employed is one that was originally developed in [Wong91]; it results in a factor of two reduction in hardware complexity, and is of sufficient merit that it should be mentioned here. The idea is to utilize a bandpass pulse shape instead of a lowpass one, with the bandpass centered at 1/4 of the oversampling frequency f_s. The reason for this choice of center frequency is clear: a digital $f_s/4$ sinusoid sampled at f_s has a waveform that consists entirely of +/-1, or zero. The modulation can be folded into the FIR filter by simply multiplying the filter impulse response by the sinusoid; this is a low-pass-to-bandpass transformation (Figure 4.13). By taking advantage of the fact that every other sample of the sinusoid is zero, half of the filter coefficients can be eliminated, resulting in a 2x reduction in hardware. In particular, given the parallel structure developed for the filter, two of the subfilters do not need to be implemented. The in-phase path will be modulated by a 64 MHz cosine wave, eliminating subfilters 2 and 4, and the quadrature path modulated by a 64 MHz sine, eliminating subfilters 1 and 3. In comparison to Figure 4.10b, the overall complex-valued, bandpass filter structure is shown in Figure 4.14, and the coefficients are listed in Table 4.1.

Filter 1	Filter 2	Filter 3	Filter 4
0.00513	0.00635	0.0	-0.0137
-0.0264	-0.0244	0.0	0.0400
0.0723	0.0635	0.0	-0.0977
-0.172	-0.157	0.0	0.285
0.623	0.890	1.0	0.890
0.623	0.285	0.0	-0.157
-0.172	-0.0977	0.0	0.0635
0.0723	0.0400	0.0	-0.0244
-0.0264	-0.0137	0.0	0.00635
0.00513		0.0	

Table 4.1: Quantized Filter Coefficients

The ramifications on the overall system design of this $f_s/4$ modulation need to be considered carefully. A conventional homodyne upconversion strategy simply multiplies the in-phase path by a cosine at the carrier $f_c = 1.088$ GHz, and the quadrature by a sine at the same frequency. Effectively, by introducing a digital modulation of $f_s/4$, a two-stage upconversion has been created, with the first stage at 64 MHz and the second stage at 1.024 GHz. The advantages of this are numerous:

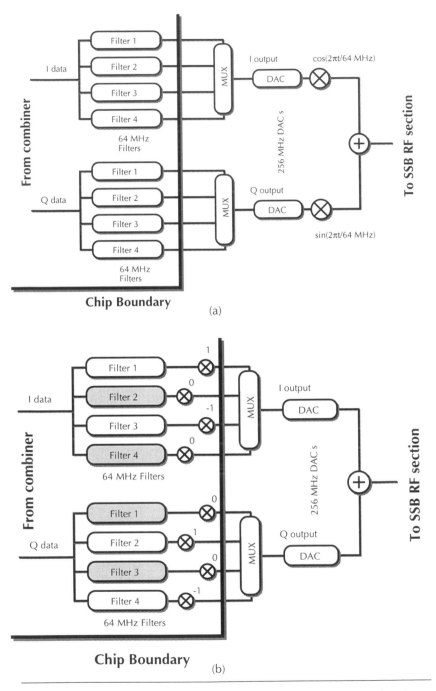

Figure 4.14: (a) Premodulation to 64 MHz (equivalent to a lowpass-bandpass pulse shape transform); (b) Folding the multiplication into the 4 parallel filters: filter complexity is halved (grey filters are unnecessary).

- Reduction in digital filter complexity by a factor of 2.

- The digital data from the chip has no spectral energy at DC, being centered at 64 MHz. AC coupling can be employed in all of the analog stages that come after the modulator, especially minimizing factors such as DC offset and low frequency noise in the subsequent DAC.

- As in superheterodyne receivers, a two-stage upconversion implies that the second oscillator will be out-of-band with respect to the transmit signal (1.024 GHz versus 1.088 GHz). Carrier feedthrough performance of the upconversion mixers has been reduced, since it will be filtered to some extent by the post-mixer spur suppression filter.

The problem that arises is the need to perform single-sideband modulation – simply multiplying by a cosine at 1.024 GHz is not sufficient, since both the positive and negative frequency images will be upconverted. The desired signal centered at 1.088 GHz will be synthesized; however, an unwanted image at 960 MHz will also be present; it is this negative frequency image that needs to be eliminated, though the use of single-sideband modulation. Chapter 5 will discuss the ramifications of this, in terms of hardware complexity of the D/A converter and modulator sections.

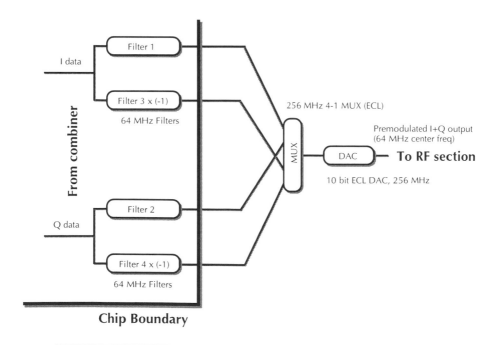

Figure 4.15: Final minimized filter architecture. One MUX/DAC can be eliminated due to the interleaved zero values in the addition.

4.2. Results

The baseband modulator circuit has been fabricated in a standard digital single-poly triple-metal 0.8 micron Hewlett-Packard CMOS process; the die photo is shown in Figure 4.15. Relevant measured die parameters are also listed there; the die has been tested to its full clock frequency of 64 MHz. The outputs, as designed, are effectively at 256 MHz, interleaved by a factor of 4. In Figure 4.16a, the spectrum of the filter output is shown, for just the pilot tone as input. Since the pilot tone spectrum is white, the output will spectrally be proportional to the filter shape; it achieves the 40 dB out-of-band rejection specification, while maintaining a 30% excess bandwidth pulse shape. The reconstructed time-domain eye diagram is shown in Figure 4.16b; it achieves perfect opening at the sample point, satisfying the Nyquist zero-ISI criteria.

One key result is the fact that this single die can support 15 users simultaneously, while consuming only 330 mW of power from a 3.3V rail. This includes all spectrum spreading, pulse shaping, combining, and per-user power control; the benefits of multiple access via digital coding are apparent here.

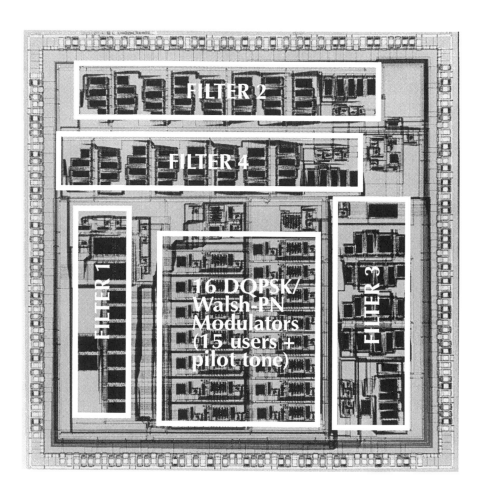

Transmit Modulator Chip	
Area:	9.9mm x 10.0mm
Transistor Count:	112000
Clock rate:	64 MHz
Power/Supply:	330mW @ 3.3V

Figure 4.16: Modulator die photo plus performance summary

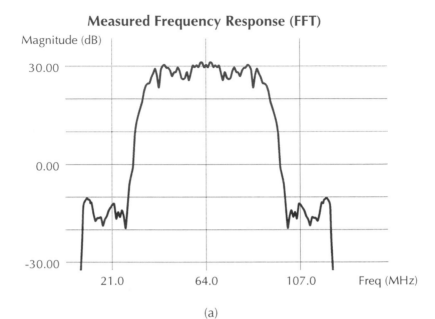

(a)

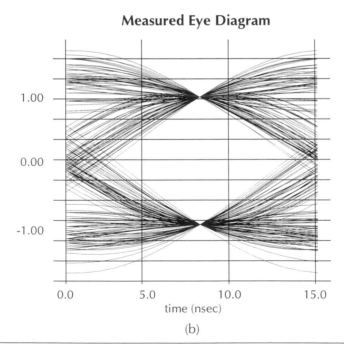

(b)

Figure 4.17: Measured modulator output. (a) FFT of die output (pilot tone only; output proportional to pulse shape); (b) reconstructed eye diagram

REFERENCES

[Beau84]   K. G. Beauchamp. *Applications of Walsh and Related Functions.* Orlando:
                Academic Press, 1984.

[Lee88]   E.A. Lee and D.G. Messerschmitt. *Digital Communication.* New York:
                Kluwer Academic Publishers, 1988.

[Peroul94]   J. Peroulas. *Design and implementation of a High-Speed CDMA Modulator for the
                INFOPAD Basestation.* M.S. Thesis, U.C. Berkeley, Berkeley, CA, 1996.

[Wong91]   B. Wong and H. Samueli. "A 200 MHz All-Digital QAM Modulator and Demodulator
                in 1.2um CMOS for Digital Radio Applications." *IEEE Journal of Solid-State Circuits,*
                Vol. 26, No. 12, pp. 1970-1980. December 1991.

[Yee96]   D. Yee. *The Design and Implementation of a Semi-Custom Transmitter for a CDMA Direct
                Sequence Spread-Spectrum Transceiver,* M.S. Thesis, U.C. Berkeley,
                Berkeley, CA, December 1996.

[Yun95]   L. Yun, D. Messerschmitt. "Variable QOS in CDMA Systems." *Proceedings of the IEEE
                ICC,* Seattle, WA. June 18-22, 1995.

5 Broadband RF Transmission and Modulation

by Dennis Yee

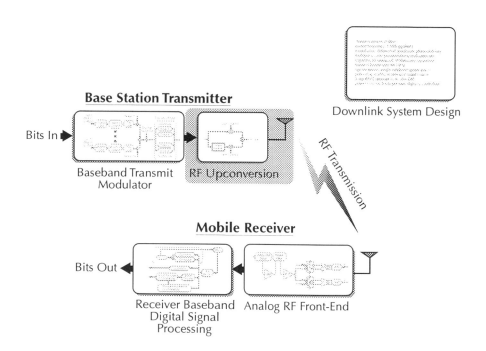

Base Station Transmitter

Bits In

Baseband Transmit Modulator

RF Upconversion

Downlink System Design

RF Transmission

Mobile Receiver

Bits Out

Receiver Baseband Digital Signal Processing

Analog RF Front-End

After spectrum spreading and pulse shaping, the digital outputs from the baseband modulator need to be converted into an analog waveform, and upconverted to the RF passband for transmission. Specifically, the 4 parallel outputs of the modulator integrated circuit must be multiplexed back into a 256 MHz serial stream, passed through a state-of-the-art 10 bit ECL D/A converter, and then modulated to a carrier frequency by a 1024 MHz carrier. In this chapter, several architectures to perform the required analog signal processing are described, especially with regards to achieving broadband, single-

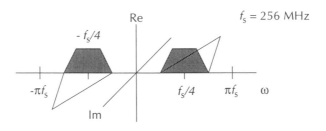

Figure 5.1: Frequency response of multiplexed baseband modulator output

sideband frequency upconversion. Single sideband is necessary since the first stage of up-conversion has been performed in the digital domain – the output of the modulator is centered at 64 MHz, not DC.

The frequency response of the 256 MHz data stream is illustrated in Figure 5.1. The in-phase and quadrature-phase data are modulated at a digital IF of 64 MHz, and may be represented by the following:

$$Y_{IF}(t) = I(t)\cos 2\pi(64\,\mathrm{MHz})t + Q(t)\sin 2\pi(64\,\mathrm{MHz})t$$

The spectrum of the desired signal for transmission at the RF of 1.088 GHz appears in Figure 5.2. Simple upconversion of the 64 MHz IF signal by multiplying by a sinusoid at 1.024 GHz (the difference between 1.088 GHz and 64 MHz) results in an incorrect spectrum at RF due to the presence of lower sidebands at 960 MHz and -960 MHz:

$$Y_{RF}(t) = [I(t)\cos 2\pi f_{IF}t + Q(t)\sin 2\pi f_{IF}t] \times \cos 2\pi f_{LO}t$$

$$f_{IF} = 64\mathrm{MHz}, f_{LO} = 1024\mathrm{MHz}$$

$$Y_{RF}(t) = \underline{\frac{1}{2}[I(t)\cos 2\pi(f_{LO} + f_{IF})t + Q(t)\sin 2\pi(f_c + f_{LO})t]} +$$

desired signal

$$\underline{\frac{1}{2}[I(t)\cos 2\pi(f_{LO} - f_{IF})t + Q(t)\sin 2\pi(f_{LO} - f_{IF})t]}$$

undesired lower sideband

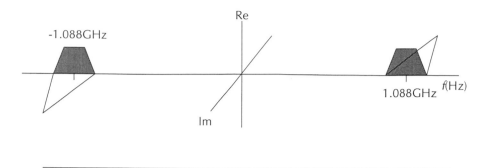

Figure 5.2: Frequency response of desired RF output signal

This negative image must be eliminated before transmission, since it represents a high-power, out-of-band signal; thus, single-sideband upconversion is necessary. Three possible methods of single-sideband upconversion are listed below:

- filter method
- phase-shift method
- Weaver method.

The advantages and disadvantages of each method are discussed.

5.1. Single-Sideband Upconversion

5.1.1. Filter Method

The filter method is the simplest of the three techniques. The block diagram of one possible implementation is illustrated in Figure 5.3. This method requires only a single mixer and bandpass filter. In accordance with the 30% excess bandwidth of the baseband pulse shape, the required bandpass filter has a passband of 83.2 MHz centered at 1.088 GHz. In addition, to meet the desired spurious out-of-band rejection, greater than 40 dB of attenuation is required within a 44.8 MHz transition band, which is about 4% of the filter center frequency.

Several different types of bandpass filters exist. Tunable LC filters are attractive since they have center frequencies which may be adjusted for use with different carrier frequencies. However, the achievable stopband attenuation of LC filters with center frequencies in excess of 1 GHz falls well short of the desired 40 dB rejection requirement. Mechanical resonance filters, such as ceramic resonators and SAW filters, offer high stopband rejection but are restricted to center frequencies under 1 GHz. Finally, electrical resonance filters offer the high selectivity of mechanical resonance filters but also operate at center frequencies in excess of 1 GHz. The characteristics of a custom eight-pole combline filter which meets the desired bandpass specifications for this application are given in Table 5.1. The use of electrical resonance filters is impractical primarily because of the

center frequency	1.088 GHz
passband (-0.1dB ripple bandwidth)	1.046 GHz - 1.130 GHz
attenuation at 1.0016 GHz	44 dB
attenuation at 1.1744 GHz	53 dB
insertion loss	0.6 dB
dimensions	2.1 in x 1.5 in x 8.85 in

Table 5.1: Combline filter characteristics

physical size of filter capable of rejecting the nearby undesired signal: the dimensions of the combline filter mentioned above are 2.1 in by 1.5 in by 8.85 in. Thus, although the filter method is the most simple design, its implementation is impractical.

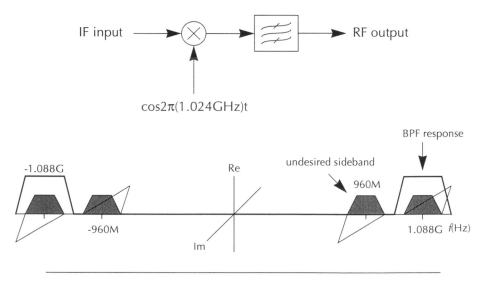

Figure 5.3: Filter method

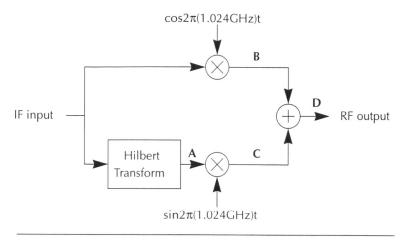

Figure 5.4: Phase-shift method

5.1.2. Phase-Shift Method

The phase-shift method is another way to achieve single-sideband upconversion. A block diagram of this method is illustrated in Figure 5.4. This method requires two mixers, an RF power combiner, and a Hilbert transformer, which can be defined as follows:

$$H(\omega) = \left\{ \begin{array}{ll} j, & \omega > 0 \\ -j, & \omega < 0 \end{array} \right.$$

The signal spectra at points A, B, C, and D in Figure 5.4 are illustrated in Figure 5.5. This method of single-sideband upconversion relies on the cancellation of the lower sideband. However, perfect cancellation is impossible in practical implementations due to amplitude and phase mismatches in the two mixers as well as in the Hilbert transformer. The input signal may be expressed as

$$I(t)\cos[2\pi(64MHz)t] + Q(t)\sin[2\pi(64MHz)t]$$

If a practical implementation of the Hilbert transformer results in a maximum amplitude difference from unity of δ_H and a maximum phase difference from 90° of ϕ_H, then the output of the Hilbert transformer may be expressed as

$$(1 + \delta_H)\{-I(t)\sin[2\pi(64MHz)t + \phi_H] + Q(t)\cos[2\pi(64MHz)t + \phi_H]\}$$

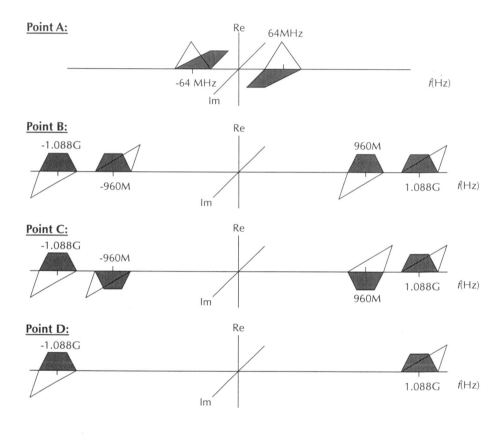

Figure 5.5: Frequency response at various points for the phase-shift method

If the maximum amplitude mismatch in the two mixers is δ_M and the maximum phase mismatch from quadrature is ϕ_M, then the output signal from the in-phase mixer is:

$$\frac{1}{2}\{ I(t)\cos[2\pi(1088\,MHz)t] + Q(t)\sin[2\pi(1088\,MHz)t] +$$

$$I(t)\cos[2\pi(960\,MHz)t] + Q(t)\sin[2\pi(960\,MHz)t] \}$$

and the output from the quadrature-phase mixer is given by:

$$\frac{1}{2}(1+\delta_H)(1+\delta_M)\{ \cos[2\pi(1088\,MHz)t][I(t)\cos(\phi_H + \phi_M) + Q(t)\sin(\phi_H + \phi_M)] +$$

$$\sin[2\pi(1088\,MHz)t][Q(t)\cos(\phi_H + \phi_M) - I(t)\sin(\phi_H + \phi_M)] -$$

$$\cos[2\pi(960\,MHz)t][I(t)\cos(\phi_H - \phi_M) + Q(t)\sin(\phi_H - \phi_M)] -$$

$$\sin[2\pi(960\,MHz)t][Q(t)\cos(\phi_H - \phi_M) - I(t)\sin(\phi_H - \phi_M)]\}$$

Thus, after the outputs of the in-phase and quadrature-phase mixers are combined, residual energy in the sidelobe centered at 960 MHz remains. In addition, interference from the quadrature-phase component appears in the in-phase sidelobe centered at 1.088 GHz and interference from the in-phase component appears in the quadrature-phase sidelobe centered at 1.088 GHz. A typical value of amplitude imbalance for a monolithic QPSK modulator is ±1.0 dB [Hewlett-Packard93] and a typical value of phase imbalance for a quadrature clock generator is ±2.0 ° [Mini-Circuits92]. A typical 90° power splitter exhibits ±1.2 dB of amplitude imbalance and ±2.0 ° of phase imbalance [Mini-Circuits92]. For a voltage swing of 1V, $|\delta_H|$=130 mV, $|\phi_H|$=2.0°, $|\delta_M|$=100 mV, and $|\phi_M|$=2.0°, resulting in about 20 dB of sideband rejection. It is possible to increase the sideband rejection to about 35 dB by adjusting the amplitude of the in-phase and quadrature-phase input signals to the QPSK modulator. Additional sideband rejection may be obtained by using a bandpass filter after the RF power combiner.

The most difficult aspect of using the phase-shift method for this application is implementation of a Hilbert transformer which operates between 22.4 MHz and 105.6 MHz (130% of the center frequency) with a phase imbalance no greater than 2.0°. Commercially available 90° power splitters which meet this specification do not exist. The next two sections discuss custom implementations of the Hilbert transformer. The first section focuses on analog implementations, whereas the second discusses digital ones.

5.1.2.1. Hilbert Transformer: Analog Implementations

The main advantage of performing the Hilbert transform in the analog domain is that only one digital-to-analog converter is required for the entire transmitter. However, the realization of a Hilbert transformer with constant amplitude and phase outputs from 22.4 MHz to 105.6 MHz is highly nontrivial.

RLC Phase-Splitter Circuit

One passive network which produces in-phase and quadrature-phase outputs is depicted in Figure 5.6 [Luck49]. Selecting $\omega_1=2\pi(22.4 \text{ MHz})$ and $\omega_2=2\pi(105.6 \text{ MHz})$ gives

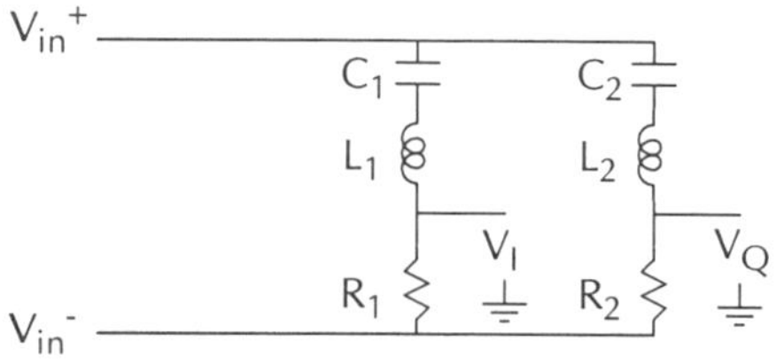

$$\omega_1^2 = \frac{\omega_0^2}{r^2} = \frac{1}{L_1 C_1}$$

$$\omega_2^2 = \omega_0^2 r^2 = \frac{1}{L_2 C_2}$$

$$Q^2 = \frac{L_1}{C_1 R_1^2} = \frac{L_2}{C_2 R_2^2}$$

Figure 5.6: RLC phase-splitter circuit

$\omega_0=2\pi(48.6 \text{ MHz})$ and $r=2.17$. The phase difference ψ between V_I and V_Q is related to the frequency ω through the following equation:

$$\tan\frac{1}{2}\psi = \frac{\frac{1}{Q}\left(r - \frac{1}{r}\right)\left(\frac{\omega}{\omega_0} + \frac{\omega_0}{\omega}\right)}{\frac{1}{Q^2} - \left(r - \frac{1}{r}\right)^2 - 4 + \left(\frac{\omega}{\omega_0} + \frac{\omega_0}{\omega}\right)^2}$$

Selecting the phase difference to be 90° at $\omega=\omega_0$ results in the following expression for Q as a function of r:

$$Q = \frac{\sqrt{2} - 1}{r - \frac{1}{r}}$$

The corresponding value for Q is 0.242, and plot of ψ as a function of f is given in Figure 5.7. The maximum phase variation from 90° for this case is about 5°, which is too high if the desired sideband rejection of 40 dB is to be achieved. The phase variation over the frequency band of interest may be decreased by using a different set of values for f_0, r, and Q. A plot of ψ as a function of frequency for $f_0=48.6 \text{ MHz}$, $r=1.9$, and $Q=0.302$ appears in Figure 5.8, in which the phase variation has been significantly reduced.

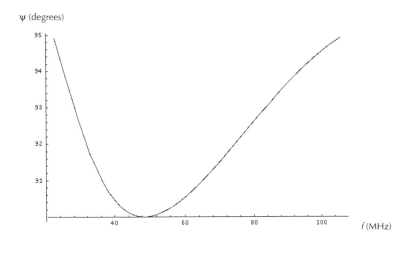

Figure 5.7: RLC phase-splitter ψ response (f_0=48.6 MHz, r=2.17, Q=0.242)

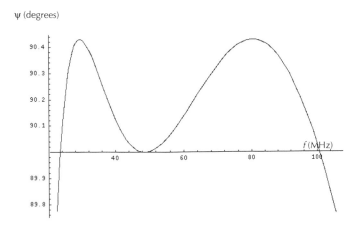

Figure 5.8: RLC phase-splitter ψ response (f_0=48.6 MHz, r=1.9, Q=0.302)

The maximum phase variation from 90° for this case is only about 1°. Although this implementation of the Hilbert transformer appears very attractive due to the theoretical phase imbalance of only 1°, two major drawbacks exist: non-constant phase relationship of individual outputs; and component value variation and mismatch. The phase-frequency dependence for ϕ_I and ϕ_Q for f_0=48.6 MHz, r=1.9, and Q=0.302 is plotted in Figure 5.9. Although ϕ_Q is offset from ϕ_I by 90° across the entire band of interest, the individual phase curves are not constant, resulting in non-constant phase offsets across the entire frequency band. The magnitudes of V_I and V_Q are equal to half the input voltage magnitude across the entire frequency band.

The implementation of this method is limited to the use of discrete components as a monolithic approach is not viable. If a 100 nH surface mount inductor with a 5% tolerance is used for L_1 and L_2, then the following component values are required for f_0=48.6 MHz, r=1.9, and Q=0.302: R_1=53.2Ω, R_2=192.1Ω, C_1=387 pF, and C_2=29.7 pF. SPICE simulation using the above component values results in the plots shown in Figure 5.10.

The SPICE results agree well with predicted performance. However, the component values used for this simulation are not commercially available, and in addition, this simulation does not take into account component tolerances. Minimization of parasitic resistance, capacitances and inductances requires the use of surface mount components with very small form factors, such as obtained in a 0603 package. Resistors with 1% tolerance are commercially available with values 53.6Ω and 191Ω [Digi-Key96]. Capacitors with 10% tolerance are commercially available with values 390 pF and 27 pF [Digi-Key96]. SPICE simulation of the RLC phase-splitter circuit using commercially available values results in

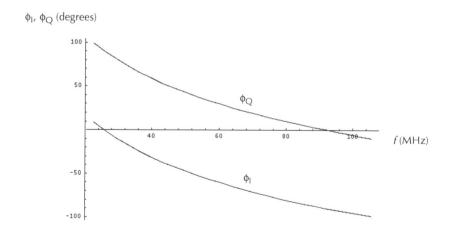

Figure 5.9: RLC phase-splitter ϕ_I and ϕ_Q response (f_0=48.6 MHz, r=1.9, Q=0.302)

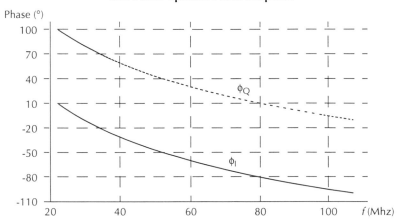

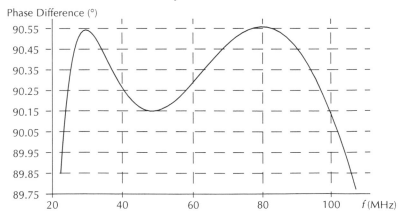

Figure 5.10: RLC phase-splitter simulated response
(R_1=53.2Ω, R_2=192.1Ω, C_1=387pF, C_2=29.7pF, L_1=100nH, L_2=100nH)

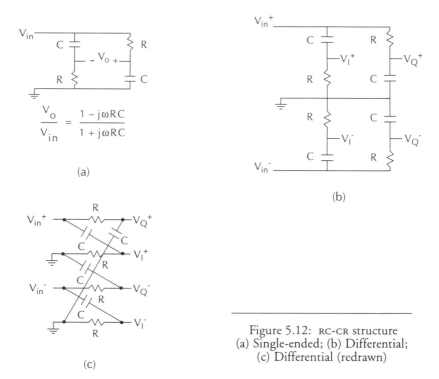

$$\frac{V_o}{V_{in}} = \frac{1 - j\omega RC}{1 + j\omega RC}$$

(a)

(b)

(c)

Figure 5.12: RC-CR structure
(a) Single-ended; (b) Differential;
(c) Differential (redrawn)

the plots shown in Figure 5.11. For this case, the maximum phase variation from quadrature is about 7°. Thus, the RLC phase-splitter network is very sensitive to component value deviations and is not amenable to discrete realizations.

Asymmetric Polyphase Networks

The most simple asymmetric polyphase network is the RC-CR structure depicted in Figure 5.12a. The voltage gain is unity for all frequencies and the phase of V_o is offset from that of V_{in} by 90° at $\omega=1/RC$. A differential version of the RC-CR structure is depicted in Figure 5.12b. The differential output voltage $V_Q=V_Q^+-V_Q^-$ is offset from the differential output voltage $V_I=V_I^+-V_I^-$ by 90° at all frequencies and the amplitude of both output voltages is $1/\sqrt{2}$ times that of the input voltage at $\omega=1/RC$. The results of a SPICE simulation for R=4.97 kΩ and C=0.5 pF are depicted in Figure 5.13.

As seen in the SPICE results, perfect quadrature is maintained across the entire frequency band. However, the phase response of the individual outputs is not constant. SPICE results also confirm that the gain is equal to $1/\sqrt{2}$ at $\omega=1/RC$ and for a 1V input the maximum output amplitude mismatch is about 600 mV. Figure 5.12c depicts a redrawn version of

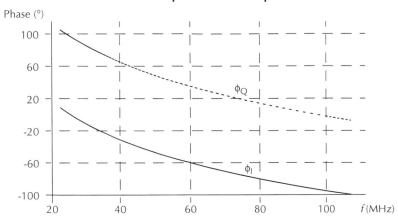

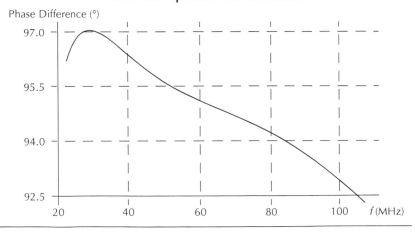

Figure 5.11: RLC phase-splitter simulated response
using commercially available component values
(R_1=53.6Ω, R_2=191Ω, C_1=390pF, C_2=27pF, L_1=100nH, L_2=100nH)

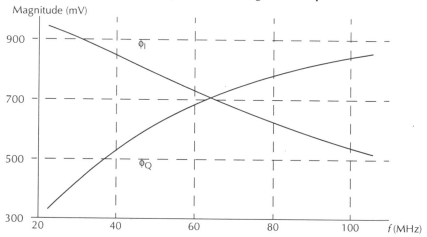

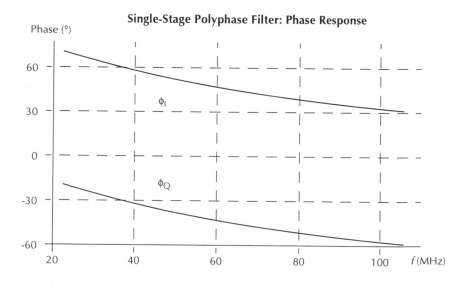

Figure 5.13: SPICE simulation of single-stage asymmetric polyphase filter
(R=4.97kΩ, C=0.5pF)

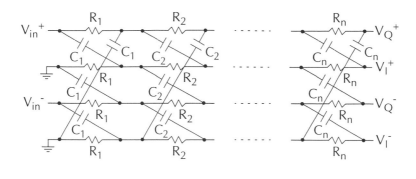

Figure 5.14: N-stage asymmetric polyphase filter

the same differential RC-CR network shown in Figure 5.12b. A major disadvantage of the differential RC-CR polyphase network is that output voltage amplitudes are equal at only one frequency. One solution is to cascade multiple stages as depicted in Figure 5.14 [Gingell73, West91, Crols95].

Each stage contributes a pole at $s=1/R_xC_x$ and a zero at $s=-1/R_xC_x$. SPICE simulation for a two-stage asymmetric polyphase filter with $R_1=R_2=4.97$ kΩ and $C_1=C_2=0.5$ pF results in the plots depicted in Figure 5.15. Again, perfect quadrature is maintained across the entire frequency band. However, for a 1V input voltage, the maximum output amplitude mismatch is now about 250 mV. This mismatch may be decreased further by cascading additional filter stages. However, the insertion loss is directly proportional to the number of cascaded stages, and thus, the output signals may require amplification if many stages are used. Another way to decrease the amplitude mismatch is by selecting different values for ω_1 and ω_2. The results of a SPICE simulation for a two-stage asymmetric polyphase filter with $R_1=14.21$ kΩ, $R_2=3.01$ kΩ, and $C_1=C_2=0.5$ pF are depicted in Figure 5.16. For this case, the maximum amplitude mismatch is about 100 mV. The best amplitude mismatch is achieved by using multiple stages and selecting $\omega_x=1/R_xC_x$ of each stage to be a different value. Although perfect quadrature is maintained over the entire frequency band, the individual phase response of each output varies as much as 40° over about 85 MHz. Due to the non-constant phase response of these asymmetric polyphase filters, their utility is limited to relatively narrow-band applications, such as generating in-phase and quadrature-phase local oscillator signals.

5.1.2.2. Hilbert Transformer: Digital Implementations

As seen above, analog methods of Hilbert transformation are hardly ideal; they are sensitive to small changes in component values, and limited in bandwidth. However, it is possible to implement the Hilbert transform in the digital domain. The primary

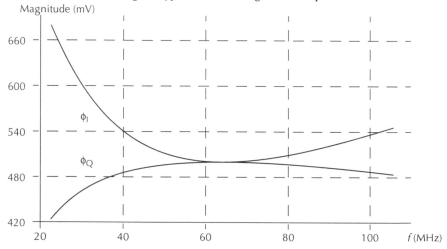

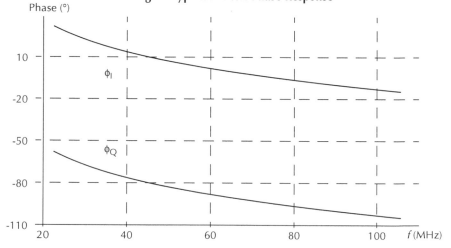

Figure 5.15: SPICE simulation of two-stage asymmetric polyphase filter
(R_1=R_2=4.97kΩ, C_1=C_2=0.5pF)

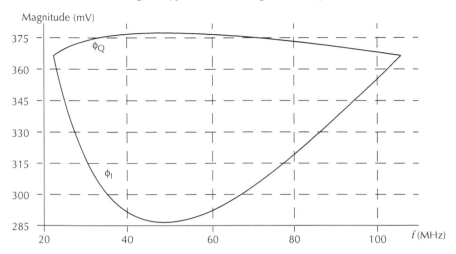

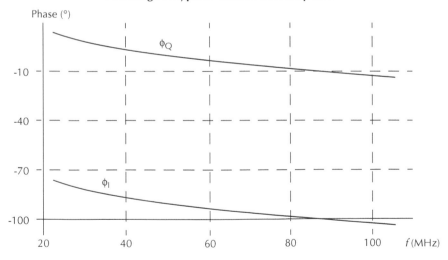

Figure 5.16: SPICE simulation of two-stage asymmetric polyphase filter
(R_1=14.21kΩ, R_2=3.01kΩ, C_1=C_2=0.5pF)

disadvantage of a digital Hilbert transform is that two digital-to-analog converters are required for the complete transmitter circuitry. However, since the transmitter resides in the basestation, the additional power consumption is not critical.

Post-Modulator Double-Sideband to Single-Sideband Converter

Multiplexing the in-phase and quadrature-phase output data streams from the baseband modulator results in a single 256 MHz digital data stream, and thus, a 256 MHz digital Hilbert transformer is required. Digital Hilbert transformers have been designed which operate at frequencies as high as 300 MHz [Hawley95].

The frequency response of an ideal Hilbert transformer is given by

$$H(\omega) = e^{-j\left(\frac{\pi}{2}\right)\text{sgn}(\omega)}$$

and the corresponding time domain impulse response is given by

$$h[n] = \begin{cases} 0 & n = 0 \\ \dfrac{\sin^2(\pi n / 2)}{\pi n / 2} & n \neq 0 \end{cases}$$

The ideal transform is not physically realizable since it is noncausal and of infinite duration. A practical realization requires windowing and shifting the sequence, resulting in a finite and causal impulse response. However, such an implementation results in finite stopband energy as well as passband ripple. The number of taps as well as coefficient quantization and internal number representation must be optimized in order to minimize passband ripple and energy in the stopband.

One way of interpreting the output of the Hilbert transformer is that it is a sequence similar to the input but phase shifted by 90°. For example, the in-phase data is multiplied by $\cos[2\pi n/4]$ in the baseband modulator chip. Since $\cos[2\pi n/4]$ is equal to zero for all odd values of n, these values of the in-phase data need not be computed and are discarded. The function of the Hilbert transformer is to determine the values at these discarded points. This interpretation of the output of the Hilbert transformer suggests a more efficient way of performing single-sideband upconversion, which will be discussed below.

Incorporation of a Digital Hilbert Transformer within the Custom Baseband Modulator

The baseband modulator chip incorporates two important design simplifications. First, since both in-phase and quadrature-phase data sequences are oversampled by a factor of four, each corresponding 256 MHz raised-cosine filter may be divided into four parallel subfilters each operating at 64 MHz. Second, the in-phase and quadrature-phase data streams are modulated by $\cos[2\pi n/4]$ and $\sin[2\pi n/4]$, respectively. Since all odd samples of $\cos[2\pi n/4]$ are equal to zero and all even samples of $\sin[2\pi n/4]$ are equal to zero, the corresponding subfilters may be eliminated, and thus, only two parallel filters each operating at 64 MHz are required for each 256 MHz raised-cosine filter.

The outputs of the baseband modulator chip are $I[n]\cos[2\pi n/4]$ and $Q[n]\sin[2\pi n/4]$. The Hilbert transforms of these two signal streams are $I[n]\sin[2\pi n/4]$ and $-Q[n]\cos[2\pi n/4]$. Thus, the function of the post-modulator digital Hilbert transform is to reconstruct the samples of the in-phase and quadrature-phase data which have been discarded. Instead of reconstructing the values of $I[n]\sin[2\pi n/4]$ and $-Q[n]\cos[2\pi n/4]$ with a post-modulator Hilbert transformer, these values may be generated directly within the baseband modulator chip. The architecture is depicted in Figure 5.17.

The signal $I[n]\cos[2\pi n/4]+Q[n]\sin[2\pi n/4]$ along the signal path denoted by the solid lines is identical the output of the current baseband modulator chip, as illustrated in Figure 5.17. The signal $I[n]\sin[2\pi n/4]-Q[n]\cos[2\pi n/4]$ along the signal path denoted by the dashed lines is its corresponding Hilbert transform. The shaded subfilters, Filter 2 and Filter 4, used to generate the Hilbert transform output along the in-phase data path are identical to the unshaded subfilters, Filter 2 and Filter 4, along the quadrature-phase data path. Similarly, the shaded subfilters, Filter 1 and Filter 3, used to generate the Hilbert transform output along the quadrature-phase data path are identical to the unshaded subfilters, Filter 1 and Filter 3, along the in-phase data path. Thus, the procedure used to implement the subfilters in the current baseband modulator chip, which is described in [Peroulas95], may also be used to implement the additional four subfilters required to generate the Hilbert transform output. The filter coefficients are listed in Table 5.2.

Filter 1	Filter 2	Filter 3	Filter 4
0.005126953125	0.00634765625	0.0	-0.01365661621
-0.0263671875	-0.0244140625	0.0	0.0400390625
0.072265625	0.06353759765625	0.0	-0.09765625
-0.171875	-0.1572265625	0.0	0.28515625
0.623046875	0.890625	1.0	0.890625
0.623046875	0.28515625	0.0	-0.1572265625
-0.171875	-0.09765625	0.0	0.06353759765625
0.072265625	0.0400390625	0.0	-0.0244140625
-0.0263671875	-0.01365661621	0.0	0.00634765625
0.005126953125		0.0	

Table 5.2: Quantized filter coefficients

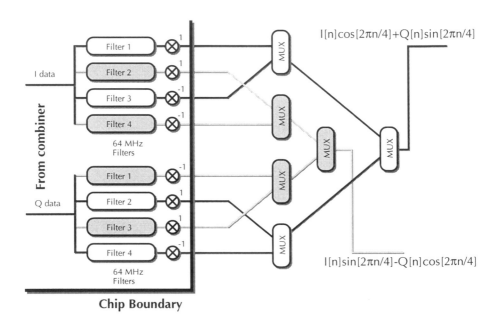

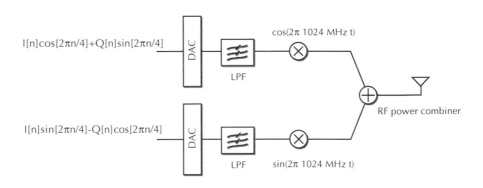

Figure 5.17: Architecture of Hilbert transform incorporated within the baseband modulator chip

This method of generating the Hilbert transform results in minimum amplitude and phase distortion at the expense of increasing the complexity of the baseband modulator chip. The area of the current baseband modulator chip, which was fabricated in a 0.8μm CMOS process, is 1x1cm^2. Since this method requires a total of eight subfilters instead of just four, incorporating the additional four subfilters onto the same chip may be impractical from an area standpoint unless the design is fabricated in a process with a smaller minimum feature size.

In addition to providing minimum amplitude and phase distortion, this method maintains the added benefit of modulating the baseband signal to an intermediate frequency (IF) of 64 MHz. Since the output data from the baseband modulator chip has no spectral energy at DC, AC coupling may be used to minimize DC offset as well as low frequency noise introduced by the digital-to-analog converters. In addition, local oscillator feedthrough in the mixers does not appear in-band, due to the 64 MHz frequency offset.

Although this method is the most attractive solution, its implementation necessitates a re-design and re-fabrication of the baseband modulator chip. Due to time constraints, this method was not pursued. Instead a method which makes use of the existing baseband modulator chip was implemented. This method is a digital variant of the Weaver method and will be discussed later. Nevertheless, future versions of the transmitter should incorporate the techniques mentioned in this section.

5.1.3. Weaver Method

Another method of single-sideband upconversion is the Weaver method. A block diagram of the Weaver method appears in Figure 5.18. This method involves two stages of mixing as well as two lowpass filters. Quadrature clock generators are required for both frequencies f_1 and f_2. For f_{IF}=64 MHz and f_c=1.088 GHz, f_1 and f_2 must satisfy the following equation:

$$f_2 - f_1 + 64\,\text{MHz} = 1088\,\text{MHz}$$

5.1.3.1. Double Upconversion Method

In traditional implementations of the Weaver method, the frequency of the first stage of mixers is selected to be at least the single-sided bandwidth of the input signal. For an IF input signal with a single-sided bandwidth of 105.6 MHz as illustrated in Figure 5.18, selecting f_1=110 MHz requires f_2=1.134 GHz and a lowpass filter with a passband from DC to 87.6 MHz and a stopband after 132.4 MHz with at least 40 dB attenuation. The frequency response at points A, B, C, D, E, F, and G are depicted in Figure 5.19. The mixer gain constraints as well as the quadrature phase requirements of this method are similar to those specified for the phase-shift method. However, one major disadvantage of this method is the introduction of an extra local oscillator (LO) frequency. The use of an extra LO frequency may be avoided by using a down-up Weaver method.

5.1.3.2. Down-Up Method

The Weaver method is traditionally used for single-sideband upconversion of baseband signals which have spectral energy down to DC. However, the input signal for this case has been digitally modulated to an IF of 64 MHz. Thus, one way to achieve the desired RF output signal is to first undo the IF modulation and then directly mix the baseband signal to 1.088 GHz. This may be achieved by using the Weaver method and selecting f_1=64 MHz and f_2=1.088 GHz. The demodulation may be efficiently performed in the digital domain by a simple ±1 multiply as depicted in Figure 5.20.

Since the baseband modulator chip operates at 64 MHz, an additional LO is not required to generate f_1. However, the demodulated signal is no longer oversampled by a factor of four, since all odd samples of the in-phase data and all even samples of the quadrature-phase data have been discarded in the baseband modulator chip. The demodulation is equivalent to downsampling the signal by a factor of two, and thus, the effective oversampling ratio of the demodulated baseband signal is just two. The downsampling process is illustrated in Figure 5.21. Due to the lower oversampling ratio, droop introduced by the digital-to-analog converter becomes more significant. A reconstruction filter may be added to compensate for the frequency droop [Lin90].

Although implementation of the down-up Weaver method requires two digital-to-analog converters, each converter operates at 128 MHz rather than at 256 MHz. Also, since the baseband in-phase and quadrature-phase signals now have spectral energy at DC, carrier feedthrough as well as DC offset become relevant issues. Despite these shortcomings, the down-up Weaver method was selected due to ease of implementation. A practical realiza-

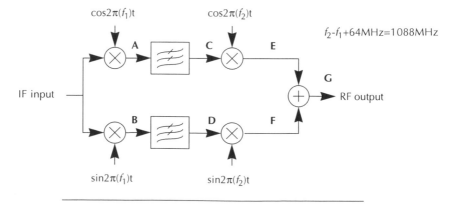

Figure 5.18: Weaver method

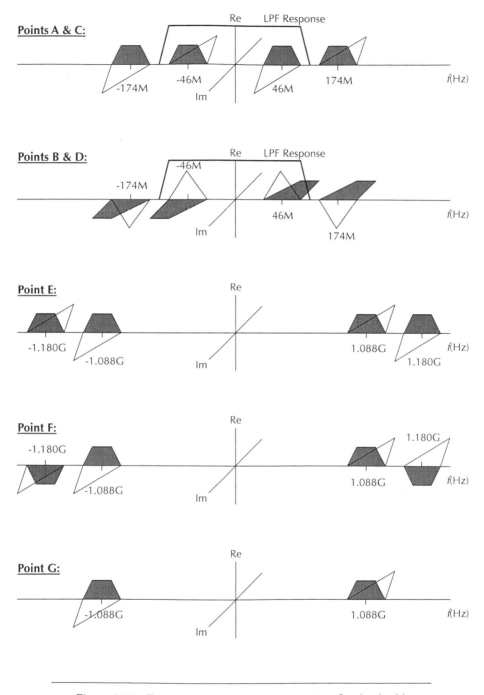

Figure 5.19: Frequency response at various points for the double
upconversion Weaver method

tion of this method requires the use of only commercially available components and the baseband modulator chip. Throughout the design process, care was taken to minimize DC offset as well as LO feedthrough.

5.2. Hardware Requirements

The hardware requirements for the implementation of the various methods have been described briefly. Table 5.3 gives more detailed information about the hardware specifications. All implementations but one make use of the existing baseband modulator chip operating at 64 MHz. The phase-shift method, which incorporates the Hilbert transform within the baseband modulator chip, requires the design of four more raised-cosine subfilters used to generate $I[n]\sin[2\pi n/4]$ and $-Q[n]\cos[2\pi n/4]$. All methods require the use of multiplexers to combine the parallel output streams from the baseband modulator. However, these multiplexers are not indicated in Table 5.3, as they are not a significant factor in determining the feasibility of an implementation. For the same reason, the inverters required for downsampling in the down-up Weaver method are also not listed.

The filter method as well as the phase-shift method employing an analog Hilbert transformer are impractical due to the need for components which cannot be easily implemented. The former method requires a bandpass filter which provides at least 40 dB

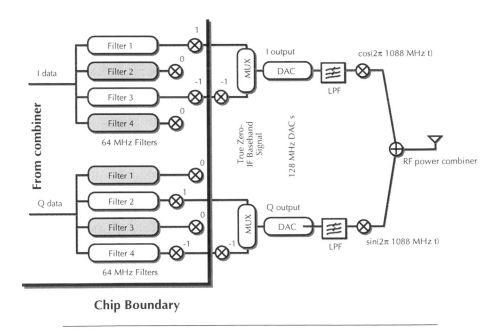

Figure 5.20: Architecture of down-up Weaver method

attenuation within the transition band 1.0016 GHz to 1.0464 GHz and the latter method requires an analog Hilbert transformer with constant amplitude and phase response from 22.4 MHz to 105.6 MHz.

The two digital phase-shift methods are almost identical in terms of hardware complexity. The method which requires a digital Hilbert filter external to the baseband modulator chip is less attractive since the Hilbert filter must operate at 256 MHz. On the other

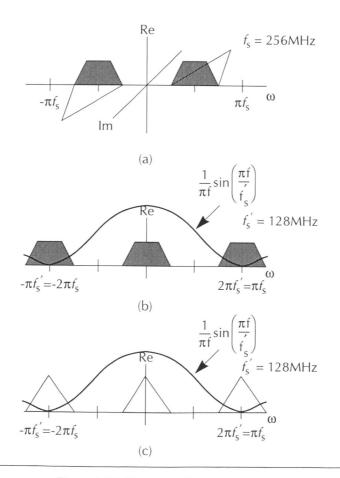

Figure 5.21: Downsampling process
(a) Output of baseband modulator; (b) Downsampled in-phase data;
(c) Downsampled quadrature-phase data

Method	10-bit DAC	Filters	RF Components	LO	Other
Filter	1@256 MHz	1 LPF (passband: DC-105.6 MHz; stopband: 150.4 MHz) 1 BPF (passband: 1.0464 GHz-1.1296 GHz; stopband: DC-1.0016 GHz)	1 Mixer	1.024 GHz	
Phase-Shift (Analog Hilbert)	1@256 MHz	1 LPF (passband: DC-105.6 MHz; stopband: 150.4 MHz)	2 Mixers; 1 Combiner	1.024 GHz (I&Q)	Analog Hilbert Transformer: 22.4 MHz-105.6 MHz
Phase-Shift (Off-Chip Digital Hilbert)	2@256 MHz	2 LPF (passband: DC-105.6 MHz; stopband: 150.4 MHz)	2 Mixers; 1 Combiner	1.024 GHz (I&Q)	Digital Hilbert Filter @256 MHz
Phase-Shift (On-Chip Digital Hilbert)	2@256 MHz	2 LPF (passband: DC-105.6 MHz; stopband: 150.4 MHz)	2 Mixers; 1 Combiner	1.024 GHz (I&Q)	Requires re-design of baseband modulator chip.
Weaver| (Double Upconversion)	1@256 MHz	1 LPF (passband: DC-105.6 MHz; stopband: 150.4 MHz) 2 LPF (passband: DC-87.6 MHz; stopband: 132.4 MHz)	4 Mixers; 1 Combiner	110 MHz (I&Q); 1.134 GHz (I&Q)	
Weaver (Down-Up)	2@128 MHz	2 LPF (passband: DC-41.6 MHz; stopband: 86.4 MHz)	2 Mixers; 1 Combiner	1.088 GHz (I&Q)	

Table 5.3: Comparison of various upconversion methods

hand, incorporation of the Hilbert transform within the baseband modulator chip requires the design of four additional sub-filters, all of which operate at 64 MHz. Nevertheless, both methods require significant time to design and fabricate the custom filters.

Finally, the down-up Weaver method is the most simple from an implementation standpoint. Although two digital-to-analog converters are required, each operates at only 128 MHz instead of 256 MHz. Also, other than the baseband modulator chip, only commercially available components are needed. As long as DC offset and local oscillator feedthrough are carefully controlled, this method is the most viable, barring the availability of a superior digital process to implement the extra filters required by the true digital Hilbert modulator architecture.

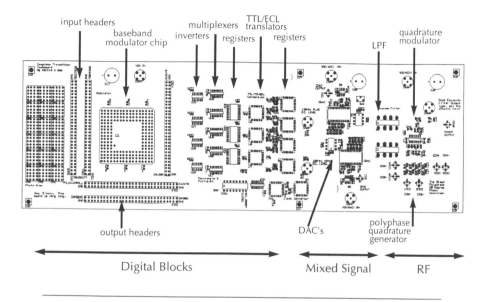

Figure 5.22: Final board layout and component placement

5.3. Results

Using the down-up single-sideband upconversion method, a semicustom board has been
developed to interface the baseband digital modulator to a pair of digital-to-analog converters, and then quadrature modulate the resulting analog signal up to the carrier frequency. The layout of the board is shown in Figure 5.22; the digital logic necessary to
achieve the down-up conversion is implemented using discrete CMOS gates, and voltage
compliance for the DAC is provided by a pair of CMOS-ECL converters. The 90° polyphase
shifting networks had an interesting application on the board: it is used not for single-
sideband modulation, but to provide a wideband means of generating the quadrature carrier signal. The reason that a wideband solution is desirable is from an experimentation
standpoint; although the system carrier is set, the ability to vary the carrier provides an
excellent means of characterizing and diagnosing performance in both the transmitter
and the receiver.

The measured spectrum of baseband analog output is shown in Figure 5.23. Due to supply bounce and clocking issues at the full 64 Mchip/sec rate (clocked at 128 MHz on the
board), measurements are shown at half-speed (32 Mchip/sec spread-spectrum rate). The
spreading factor remains at 64; however, each user's data rate is at 1 Mbps instead of
2 Mbps. The waveform corresponds precisely to that of the desired 30% excess band-

magnitude (dB)

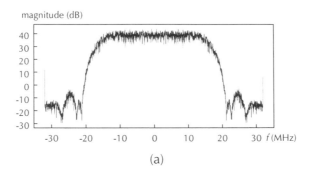

(a)

magnitude (dB)

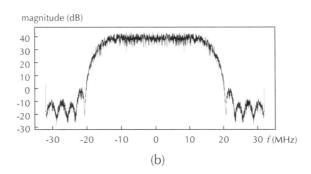

(b)

Figure 5.23: Measured analog output spectra
(a) in-phase output; (b) quadrature output

width raised cosine filter, with 40 dB of rejection at the first sidelobe. Maximum operating frequency for the RF carrier output is set by the performance of the polyphase shifter
network, which was 990 MHz; higher performance (up to 2 GHz) is achieved using an
off-board quadrature generator, limited solely by the particular on-board RF quadrature
modulator/mixer that was employed. RF functionality was verified by using the custom
analog receiver front-end to demodulate the signal and observing the received baseband
signal; these results are shown at the end of Chapter 7 as part of the receiver discussion.

The power consumption of the overall board is shown in Table 5.4. Not surprisingly, the
largest contribution to the overall board power is from the high-speed board-level logic,
especially from the ECL circuitry. Of the total 10.35W, 7W are expended in performing
some simple ECL and CMOS digital functions. Had a better fabrication technology been
available, a custom chip could have been developed to easily perform these digital inter-

face functions, at a much lower power budget. Even the DAC, consuming 2W of power, might be possible in a more advanced CMOS technology. The advantages of high levels of integration are clear: the baseband modulator, performing coding and pulse shaping for 15 users in parallel, consumes approximately 2% of the total power, whereas simple off-chip digital logic consumes 75% of the total power.

5.4. Extensions to Multiple Baseband Modulators

The key research and development effort on the transmit side has been a fully custom baseband transmit modulator circuit, a single chip capable of synthesizing a broadband spread-spectrum signal multiplexing 15 users simultaneously. The complexity benefits of using spread-spectrum are apparent: the hardware necessary to synthesize each user's transmit signal is easily implemented using basic digital blocks, as opposed to traditional frequency division systems that require one analog oscillator per user.

The final issue that remains is how multiple die can be ganged together to support more users than 15 in a single base station, if necessary. The key thing to note is that only one pilot tone should be present; if n die are to be used in parallel, one should be designated as a timing master and bears the pilot tone; the remaining $(n-1)$ should suppress pilot generation. The most obvious solution would be to take n modulator die, clocked by the same crystal oscillator, and perform a digital addition from the corresponding outputs of all n die (Figure 5.24a). Provided their reset circuitry is released at the same instant, perfect PN and Walsh synchronization is guaranteed, since the combining of the n outputs is done in the digital domain. The key limitation to this method is the issue with dynamic range in the D/A converter: since multiple die outputs are being combined pre-DAC, the fixed 10-bit limitation of the DAC will end up reducing available dynamic range to each die. At the other extreme, another solution would be to simply utilize several complete transmit boards, and power combine their 1 GHz RF outputs (Figure 5.24b). The key drawback to this in synchronizing the die between each board – for the pilot tone system to work, the boards need to be synchronized to better than 0.125 T_{chip}, or 2 nsec[1], else the mis-synchronization would be disastrous. The optimal solution is shown in Figure 5.24c; analog combining after the DAC, but before the frequency upconversion. Since the DAC's are clocked from a common frequency reference, provided careful skew control and layout are done on the board, synchronization is preserved between the n parallel signals; likewise, no dynamic range is lost in the DAC. The obvious drawback is the need for multiple DAC's; clock coupling, radiation, and noise while preserving 10-bit linearity become an increasingly challenging board design issue.

---

1. The 0.125 T_{chip} requirement arises from the worst-case delay-locked loop timing error in the receiver (Chapter 8); it also arises from the timing tolerance in the eye diagram for a 30% excess bandwidth raised cosine (Figure 3.3).

Component	Power
Baseband modulator chip	0.2W
CMOS logic	0.8W
ECL register	2.5W
TTL-to-ECL converter	3.0W
ECL-to-TTL converter	0.75W
Analog buffers	1W
D/A converters	2.0W
Quadrature modulator	0.1W
Total:	10.35W

Table 5.4: Power consumption breakdown by components

Given that a solution has been developed to support multiple users, perform pulse shaping, and convert the signal into the 1 GHz transmit band, the remaining problem, which is also the most difficult one of the entire system, lies in the implementation of the mobile receiver. From a tiny broadband received signal off the antenna, the receiver must amplify this signal, perform analog-to-digital conversion, and provide sufficient digital signal processing capability to recover and RAKE combine the multipath arrivals. The next two chapters deal with a single-chip solution for the analog RF receiver front-end, while Chapters 8 and 9 describe the inverse of the modulator chip: the digital spread-spectrum receiver.

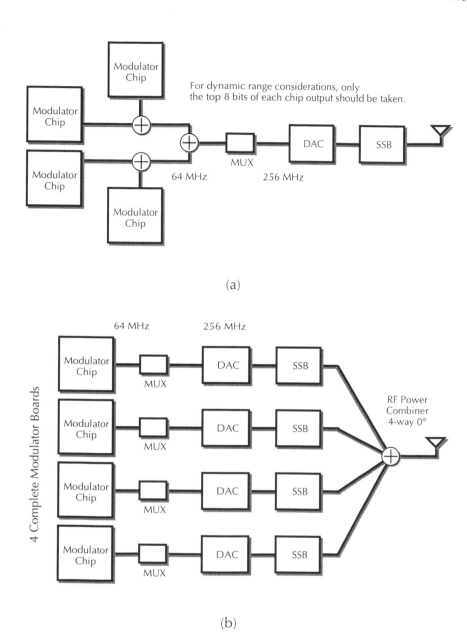

Figure 5.24: Multi-modulator combining schemes for the base station
(using the down-up method).
(a) Baseband combining of digital outputs
(b) RF combining

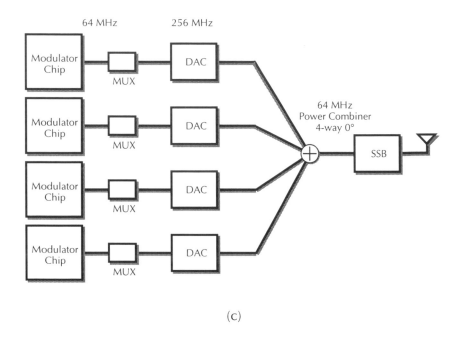

(c)

Figure 5.24, continued: Multi-modulator combining schemes for the
base station (using the down-up method).
(c) Analog baseband combining

REFERENCES

[Crols95]   J. Crols, M. Steyaert, "A Single-Chip 900MHz CMOS Receiver Front-End with a High Performance Low-IF Topology." *IEEE Journal of Solid-State Circuits*, Vol. 30, No. 12, pp. 1483-1492, December 1995.

[Digi-Key96]   *Digi-Key Catalog*. Digi-Key Corporation, 1996.

[Gingell73]   M. J. Gingell, "Single Sideband Modulation using Sequence Asymmetric Polyphase Networks," Electrical Communication, pp. 21-25, 1973.

[Hawley95]   R. A. Hawley, T. Lin, H. Samueli, "A 300MHz Digital Double-Sideband to Single-Sideband Converter in 1μm CMOS." *IEEE Journal of Solid-State Circuits*, Vol. 30, No. 1, pp. 4-10, January 1995.

[Hewlett-Packard93] *Communications Components Designer's Catalog, GaAs and Silicon Products*, Hewlett-Packard, 1993.

[Lin90]   T. Lin, H. Samueli, "A 200-MHz CMOS X/SIN(X) Digital Filter for Compensating D/A Converter Frequency Response Distortion in High-Speed Communications Systems," IEEE Global Telecommunications Conference and Exhibition, pp. 1722-1726, December 1990.

[Luck49]   D. Luck, "Properties of Some Wide-Band Phase-Splitting Networks." *Proceedings of the IRE*, pp. 147-151, 1949.

[Mini-Circuits92] *RF/IF Designer's Handbook*. Mini-Circuits Co., 1992.

[Peroulas95]   J. Peroulas. *Design and Implementation of a High Speed CDMA Modulator for the InfoPad Basestations*. M.S. Thesis, U.C. Berkeley, Berkeley, CA, 1995.

[West91]   J. West, "The Application of the Asymmetric Polyphase Filter in an SSB Transceiver," Proceedings of the South African Symposium on Communications and Signal Processing, pp. 85-92, August 1991.

6 The Receiver: Analog RF Front-End

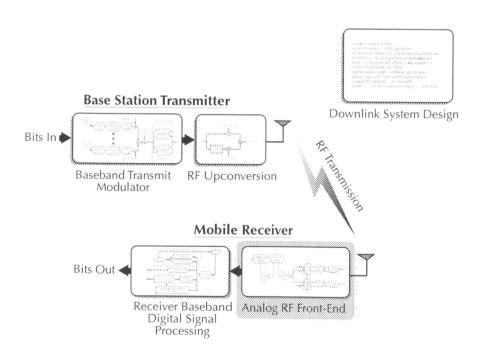

Base Station Transmitter

Bits In ►

Baseband Transmit Modulator

RF Upconversion

Downlink System Design

RF Transmission

Mobile Receiver

Bits Out ◄

Receiver Baseband Digital Signal Processing

Analog RF Front-End

The challenge presented by the mobile receiver – gigahertz-band analog performance in addition to digital signal processing at hundreds of megahertz – is the greatest barrier that needs to be surmounted if the vision of a broadband wireless terminal is to be achieved. As an architectural issue, one major design goal is that the analog hardware be simplified as much as possible; since the carrier frequencies are above 1 GHz, the complexity and difficulties in implementation imply that simplifying the circuitry or relaxing the required analog performance should be paramount. Given the tremendous levels of digital computation achievable by today's scaled MOS technologies, as much as possible of the required signal processing should be implemented at baseband, in the digital domain. Use of such techniques as sampling demodulation and homodyne receiver architectures all present new methods in developing high-performance demodulators, which take advantage of the fact that digital data – spread-spectrum digital data – is being transmitted.

Concomitant with the goal of simplification is the desire for monolithic integration of as much of the analog circuitry as possible. Traditionally, the realm of gigahertz-band RF front-end circuitry has been dominated by discrete gallium-arsenide MESFET's using stripline filters. Such discrete-element designs consume significant amount of area on a circuit board; given that separate packages need to be utilized for each component, trace lines must be placed carefully to minimize cross-coupling, and termination resistors are required to minimize transmission-line effects from the trace. Likewise, discrete designs are poor from a power consumption standpoint, since the power required to drive high-speed analog signals across interconnect lines is directly proportional to the parasitics that must be overcome. Likewise, the requirements of proper termination and impedance control are necessitated at the board level; current must be expended in driving matched 50Ω loads. Since the parasitics for on-chip interconnects are at least an order-of-magnitude lower than board-level interconnects, and short on-chip propagation distances imply that termination is unnecessary, the power savings realized by monolithic circuit integration are considerable [Bakog90]. To date, most of the applications utilizing gigahertz or higher frequencies have either been narrowband (such as voice telephony), or fixed installations (such as terrestrial microwave). In next-generation mobile systems, not only will portability be demanded, but high per-user bandwidth. Achieving the required bandwidth, along with low power consumption and small physical size, demands that highly integrated analog technologies be exploited.

6.1. Receiver Architectures

In Figure 6.1a, a conventional superheterodyne RF front-end, potentially suitable for use in this system, is shown. Schematically, anti-image filtering is performed, followed by low-noise amplification and two conversion stages of mixing and filtering. The power breakdown for this schematic is shown in Table 6.1.

This design was fabricated utilizing off-the-shelf components [Behzad95], as a reference point for what the monolithic analog front-end solution needs to achieve, and what problems it must overcome. The resulting board is shown in Figure 6.1b; it consists of slightly over 200 components, mostly off-chip biasing and passive elements, with a dozen small-scale integrated circuits to implement the active functions such as gain and mixing. Furthermore, power consumption is high: 750 mW, the vast majority of which is consumed in the RF and intermediate-frequency (IF) local oscillators, and the high-speed baseband analog-to-digital (A/D) converters.

Thus, the goals of the monolithic solution are clear: minimization of the power consumed in the receiver, and reduction from 200 components to a "single" chip. As a preview, the final architecture of the analog receiver front-end chip is shown in Figure 6.2. The entire receiver consumes 108 mW to demodulate the gigahertz-band modulated sig-

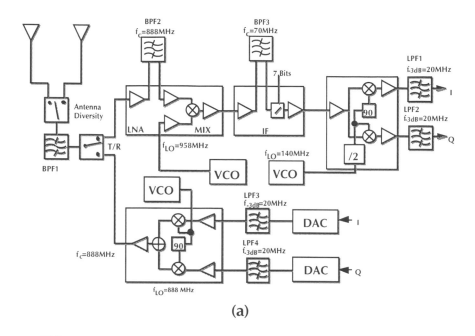

(a)

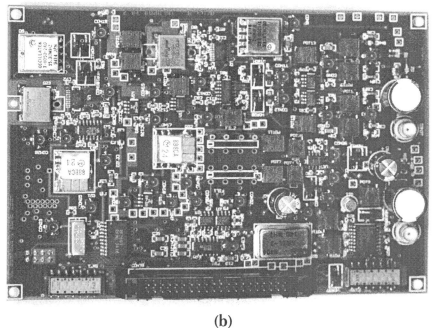

(b)

Figure 6.1: Off-the-shelf analog front-end implementation
(a) Schematic (b) Semicustom board implementation

Receiver	Power (mW)
LNA/Mixer	65
IF VGA	60
Demodulator	30
RF+IF Osc	110
Buffers	100
A/D Converter: 8b, @50MS/sec	770
Total Rx Power	1130

Transmitter	Power (mW)
Modulator	120
Buffers	45
D/A Converters: (8b, 40 MS/sec)	165
Total Tx Power	330

Table 6.1: Semicustom receiver power breakdown

nal, and is completely integrated with the exception of two image/noise rejection filters. The large power savings is achieved in two ways: reduction in the number of frequency conversion stages, i.e., the use of homodyne frequency conversion instead of heterodyne; and the development of a low-power, high-speed A/D converter. Furthermore, the single homodyne frequency conversion is not implemented with a conventional local-oscillator and mixer configuration. Instead, the RF signal is directly subsampled, which uses the harmonics of the sampling operation itself to achieve conversion. By resorting to this method of frequency conversion, power has been further reduced by the architectural elimination of several components, notably the RF and IF oscillators. In some sense, this architecture is the absolute minimum functionality an analog front-end needs to achieve: amplification, frequency translation, and A/D conversion. It depends entirely on the subsequent receiver DSP to achieve user channel selection, frequency offset compensation, and timing recovery – functions that traditionally have been done in the analog domain. To begin the discussion of this architecture, the RF aspects – low noise amplification, filtering, and homodyne sampling downconversion – will be examined in this chapter. The following chapter will examine the baseband analog issues – primarily that of high-speed A/D conversion and variable-gain amplification.

6.2. Silicon CMOS for RF Analog

Given the advances in silicon processing and technology of the past decade, digital technologies have seen a revolutionary breakthrough in both performance and size through the use of device scaling, especially in the arena of silicon complementary MOSFET's. For

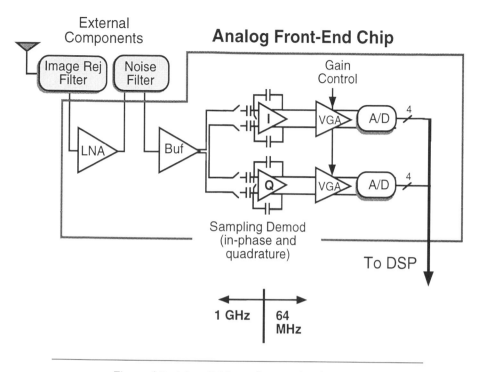

Figure 6.2: Monolithic RF front-end architecture

micro- and millimeter- wave IC (MMIC) technologies, gallium arsenide has been used almost exclusively to date, despite the extra cost and processing difficulties, simply because silicon devices have not possessed the necessary performance. However, the same benefits derived from scaling are reflected in analog applications as well; increases in digital switching speed commensurately yield increased analog bandwidths.

While MOS has been regarded in the past as a "slow" technology, the scaling of device feature sizes has increased MOS performance dramatically over the past decade, and will continue to do so as devices are reduced into the deep-submicron regime. In particular,

$$f_t = \frac{g_m}{C_{gate}} = \left(\mu_o C_{ox}\left(\frac{W}{L}\right)(V_{gs}-V_t)\right)/(W\,LC_{ox}) = \frac{\mu_o}{L^2}(V_{gs}-V_t)$$

where V_{gs} is the gate bias voltage, V_t is the device threshold voltage, C_{ox} is the oxide capacitance, and μ_o is the carrier mobility in the channel; W and L are the width and length of the device gate, respectively.

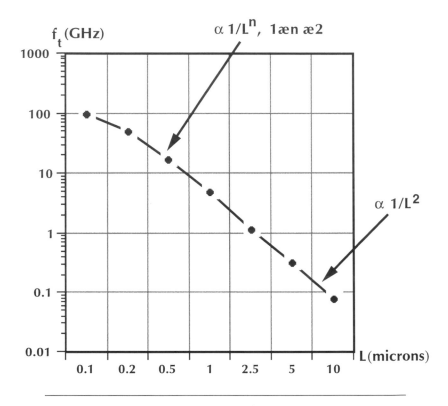

Figure 6.3: Achievable device f_t vs. MOSFET drawn L

As the channel length L is reduced by a factor λ, the f_t thus increases as λ^2. This is borne out well by the performance scaling of MOS devices seen over the past decade. In 1980, a conventional MOS device was fabricated with a minimum drawn length of 10μ, and yielded a corresponding f_t of around 80 MHz; today, a 0.8μ channel length NMOS device can achieve an f_t of 10 GHz, over 100 times larger, and certainly practical for operation at 1 GHz. In Figure 6.3, the f_t of MOS devices as a function of channel length is shown. Below 1μ, short-channel nonidealities begin to dominate, such as velocity saturation in the channel, which correspondingly degrades the transconductance g_m. Assuming complete channel velocity saturation at a carrier velocity v_{sat}, it can be shown [Watts89] that g_m and f_t are now given by $g_m = C_{ox} W v_{sat}$, and $f_t = (v_{sat} / L)$. Even under this extreme assumption, the analog device performance still scales as 1/L – yielding potentially 100 GHz f_t devices at a drawn L of 0.2μ. Given the additional benefit of reduction in parasitics, gigahertz-band monolithic CMOS circuitry is both desirable and feasible.

With the current generation of silicon technologies, the achievable f_t of conventional silicon CMOS is at performance levels required for effective operation in the low-microwave regime. In particular, with the technology available for fabrication at this time (a 0.8μ single-poly CMOS process), operation at 1 GHz can be achieved, and the choice of system carrier frequency was set by this. In future versions of this system, when non-experimental deployment needs to be considered, operation at 2.4 GHz or even 5 GHz will be required, necessitating even greater device scaling.

6.3. Low-Noise Amplification

Ostensibly, one key component of any receiver chain is the low-noise amplifier coming off the antenna; since the signal is comparatively weak, good gain and noise performance is necessary. The overall noise figure of the receiver is given as:

$$NF_{receiver} = \left(\frac{1}{G_{LNA}}\right)^2 (NF_{mixer} - 1) + NF_{LNA}$$

where NF_{mixer} is the aggregate input-referred noise figure of the subsequent amplifiers and mixing stages, and G_{LNA}, NF_{LNA} are the gain and noise figure of the LNA itself. All of the noise figures in the above expression are in linear units, not in dB. The input-referred noise of all subsequent stages is reduced by the gain of the low-noise amplifier, requiring high gain; the noise of the amplifier is injected directly into the received signal, requiring low noise. There are well-known tradeoffs in amplifiers between noise and gain; often, one is achieved at the expense of the other [Pozar90]. The inherent issues and compromises between these two will be examined below.

6.3.1. System Noise Requirements

The overall required system noise performance can be determined by examination of the transmission environment, and system simulation in the presence of additive noise. For the receiver, assuming a microcellular transmit power of 0 dBm and 0 dBi antenna gains (omnidirectional antennas), a 1 meter path-loss reference of 30 dB, and a path-loss coefficient of 2.6, the spatial-average received power at the boundary of a 4 meter cell is approximately -60 dBm; under worst-case fade conditions, a fading margin of -10 dB is added to this[1], implying that the receiver noise performance must recover a -70 dBm signal, or a zero-to-peak voltage of 0.1 mV (.07 mV RMS) across a 50Ω load.

---

1. Typical narrowband fading margins are much larger than this; however, due to the power-averaging effect of spread-spectrum across the band and the indoor Rician fading statistics, the fading margin is much smaller. Intuitively, the peaks and deep nulls are averaged out if the transmit band is much larger than the coherence bandwidth [Viter95].

Using these values, a plot of required noise figure as a function of QAM constellation size is shown in Figure 6.4, for a constant BER = 10^{-4} under the assumption that the primary noise source is due to receiver noise and not intercell interference. Two opposing factors are seen here: the decreasing bandwidth (and hence noise power) as constellation size increases, and the decreasing distance between constellation points; the first serves to lower the BER; the second increases it. Surprisingly, the peak occurs for an extremely small constellation size – for constellations larger than this the NF must drop off quickly to maintain the BER. For the broadband CDMA system in question, a 7 dB noise figure can be tolerated, which is good from the standpoint of both CMOS implementation and low power consumption, as will be shown below. In actuality, an even higher noise figure than this may be usable: the assumption that intercell interference is negligible is not true under all conditions, especially at the cell boundaries where cells overlap. In a deep fade, if the dominant noise element is from intercell interference, low-noise design in the am-

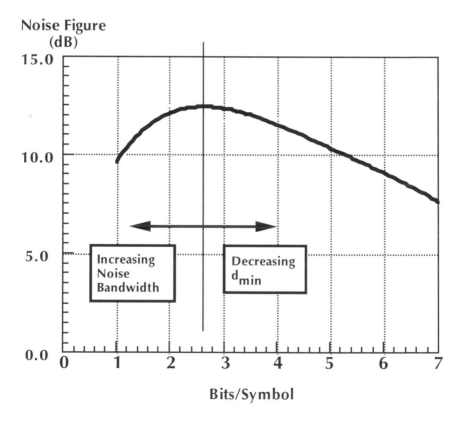

Figure 6.4: System noise figure versus constellation size, for constant BER= 10^{-4}

plifier will count for nought; error-correction coding, soft-handoff, and microcell diversity will play a far greater role in maintaining the wireless link. However, as a conservative design goal, a noise figure of 7 dB is set.

6.3.2. RF LNA Design and Matching Filter Networks

Low-noise amplifier design, and RF amplifier design in general, can be approached from one of two methods: broadband or narrowband. The broadband methodology is akin to the design of op-amps – stability, gain, and noise performance are considered across all possible frequencies, using Bode/Nyquist plots. The narrowband methodology, on the other hand, deals with the stability, gain, and noise performance of the circuit at a single frequency f_c, and the analysis is valid for a small range of frequencies around f_c– design by S-parameters and network analysis falls into this category [Gonz84]. Obviously, for the same topology, the two methods must come to the same conclusion; however, when dealing with RF tuned systems, the band of interest is narrow and centered about a carrier frequency, and hence the S-parameter method is usually of greater utility.

As far as matching is concerned, it is obvious why an input match to 50 ohms is necessary in the LNA: it will ultimately be connected to the off-chip filters and the receiver antenna. However, one immediate question that arises is why an output match to 50 ohms is necessary in the low-noise amplifier. There are several reasons for this additional requirement. First, interfacing to standard test equipment is vastly simplified, since such equipment assumes 50 ohm connections at all test ports. Second, it provides additional flexibility to the architecture; such an LNA could work in a standalone mode at 1 GHz, as opposed to being incapable of driving any off-chip load. Lastly, and most importantly, the antialiasing requirements of the sampling demodulator mixing stage (Section 6.4) demand a second filter stage after the LNA, not only to provide additional image rejection, but to filter the broadband white thermal noise of the LNA itself. For these reasons, an output 50 ohm match was a critical design aspect of the LNA topology; as will be shown below, this also proved to be one of the most difficult to meet.

The general topology of any RF amplifier can be broken down into three stages: an input matching network, an active transducer (the amplifier proper), and an output matching network (Figure 6.5). The input and output networks are passive, consisting of striplines, inductors, capacitors, and resistors; their role is to achieve optimal gain and/or noise performance in the system by transforming the source/load impedances into values that are matched to the those of the transducer. For the case of maximizing gain, maximum overall power transfer is desired. Looking back from the input of the amplifier, the input matching network should exhibit a reflection coefficient Γ_{IN} equal to S_{11}^*; i.e., the transformed source impedance should be the complex conjugate of the amplifier's input impedance (under the assumption that S_{12} is small). Likewise, the same is true for the output: Γ_{OUT} should equal S_{22}^* for maximum power transfer.

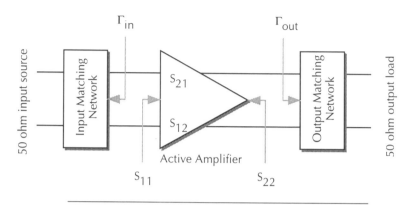

Figure 6.5: Generalized RF amplifier topology

It is useful to gain some intuition as to how the matching networks, being entirely passive, can add "gain" into the system, and why maximizing power transfer is tantamount to maximizing gain. As an example, consider an ideal transducer, with finite gain $S_{21} = 10$, an input impedance of 500 kΩ and a 50 Ω output impedance, constant for all frequencies. As shown in Figure 6.6a, this amplifier could easily be driven directly by the source impedance, resulting in an overall gain equal to S_{21}. One might mistakenly equate this with the maximum gain possible – it is not. The input is accepting effectively zero power: the S_{11} of the amplifier is equal to 1.0. In Figure 6.6b, the amplifier plus its ideal matching network is shown: a transformer with turns ratio equal to 100; looking back from the amplifier, it now sees an effective impedance equal to 500 kΩ The gain is now:

$$(S_{21} \bullet n) \frac{1}{1 + \dfrac{n^2 50\Omega}{500k\Omega}} = 500$$

representing a boost by 34 dB. The numerator is the effective voltage gain of the transformer times the transducer gain; the denominator comes from the power now being absorbed by the 500 kΩ input resistance (resistor divider with the transformed 50Ω). This is also easily seen to be the maximum gain; if the turns ratio is larger than 100, the voltage divider effect destroys any further gain. Intuitively, there are two signal-bearing variables: the voltage and the current. In the case of Figure 6.6a, no current signal is being employed; the source can provide a significant current signal, but the amplifier cannot accept it. In the case of 5.6b, the voltage and current signals are both being employed by proper adjustment of the input impedance: this is the concept of power matching.

Thus, there is a maximum achievable transducer power gain [Pozar90], which is controlled not only by the S_{21} of the amplifier, but also the S_{11} and S_{22} as well. With power matched input and output networks, it is given by:

$$\text{Gain}_{\text{transducer, max}} = \frac{1}{1 - \left|S_{11}^2\right|} \left|S_{21}\right|^2 \frac{1}{1 - \left|S_{22}^2\right|}$$

under the assumption that the amplifier exhibits good reverse isolation (S_{12} is zero). In the previous example, an ideal broadband matching element was used: a transformer. Although it is possible to utilize transformers at RF, it is difficult to achieve any reasonable

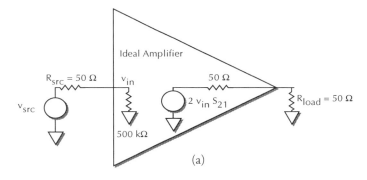

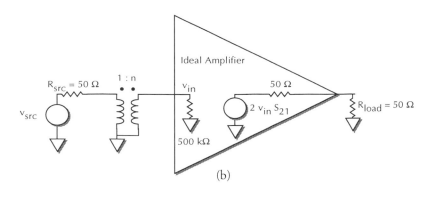

Figure 6.6: Power matching to achieve gain from passive elements.
(a) Prototype case (b) Power matched case

turns ratio, due to the interwinding capacitance limiting the frequency response of the coils. Broadband match is preferably implemented using active elements in feedback to achieve controlled impedances. However, given that the signal is inherently narrowband, another means of impedance matching is available: transformation using resonant circuits. Furthermore, narrowband resonant matching has another advantage: the required input and output noise limiting filters can be folded in as part of the matching network. Both broadband and narrowband matching techniques will be examined below; it becomes rapidly apparent that if proper impedance matching is needed at both input and output, broadband design is simply not useful from either a low-power or high-performance standpoint for the available fabrication technologies.

6.3.3. Broadband-Matched Topology

First, a low-noise RF CMOS broadband design will be examined, as a basis for a tuned design, and to explore the advantages and disadvantages inherent in such a design. A broadband design is desirable, since it can be used at any frequency in its amplification band independent of the implementation of the noise-rejection filters; the filters and amplifiers become decoupled design problems. Likewise, it should be mentioned that the original intent of this design was as a low-noise amplifier for a multichip module (MCM) [MCM91], in which 50Ω match at the output was unnecessary due to the extremely short transmission distances within the multichip and the ability to formulate filters within the multichip structure itself. Given the research issues inherent in the fabrication of the multichip, decoupling the filter and amplifier designs became critical.

The proposed amplifier topology (Figure 6.7a) is similar to what is done in conventional GaAs MMIC designs [LNA81]; it is a transconductance device (M_1) driving a tightly-coupled transresistance feedback pair (M_2-M_3). The target fabrication technology is a 1.2 micron non-double poly HP digital process, and all results below are with respect to this. Although the output resistance of the transresistance stage is low (effectively $1/g_{m3} = 1.2k\Omega$), it cannot be made sufficiently low to achieve a 50Ω match; this would require M_3 to be resized to a W/L of 300μ/1.2μ, with a higher bias current of 5 mA. Both of these are undesirable, since the higher W/L results in severe loading on the previous stage as well as vastly increasing the power consumption of the amp. Furthermore, given that the ultimate intent of the transconductance-transresistance design is for a multichip module, such matching is unnecessary as well. For testing purposes, a pair of source followers are needed to match this impedance to the 50Ω load; the output followers are removed in the multichip implementation after the core amplifier design is characterized. Two stages are used: the first buffer stage is smaller than the second, to minimize loading on the transresistance pair; the second is designed such that $1/g_m$ is equal to 50Ω

First, the DC bias of the entire topology must be considered, with all bias voltages being internally generated. The most interesting bias problem is establishing the base V_{gs} of M_1, and hence its desired drain current of 1.2 mA. Given that the RF antenna input is capacitively coupled into the amplifier at this point, it is important to prevent attenuation of the RF signal, as well as prevent the input signal from being injected into the current

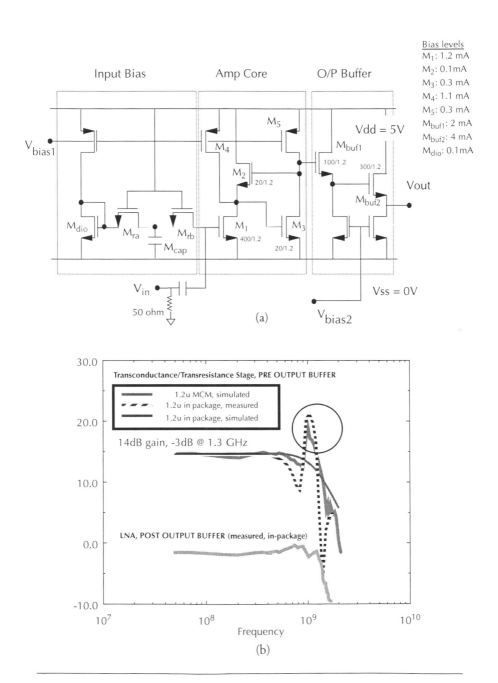

Figure 6.7: Broadband-matched LNA topology.
(a) Schematic (b) Measured response. The top 3 curves are for the transconductance/transresistance stage by itself ("MCM-mode"); the bottom curve is the amplifier with the matching output buffers.

bias chains. This is accomplished by using an R-C-R filter, formed by M_{ra}, M_{cap}, and M_{rb}. M_{ra} and M_{rb} are run in triode, with a W/L = 2.4μ/5μ; their V_{DS} is zero at DC, since no current can flow through them. Thus, they behave as resistors, with resistance equal to $1/[k'(W/L)(V_{gs}-V_t)] = 8$ kΩ . The center capacitor M_{cap} is simply a 100μ/100μ FET gate, to maximize capacitance per unit area. With respect the diode chain, the R-C-R network forms a lowpass filter, preventing any RF signal at the input from being injected into the bias references, while permitting the DC bias established by M_{dio} to be mirrored in M_1. This is the integrated equivalent of an RF bias tee, without the need for inductor chokes. To complete the biasing of the amplifier, standard cascoded PMOS current sources set up the bias in M_2 and M_3 (0.1 mA and 0.3 mA respectively); it is important to note that the current in M_2 is the difference between the M_1 bias and the current source. Non-cascoded current sources are used to bias the output source followers to 2mA and 4mA respectively, since headroom here is a serious issue.

As far as gain performance is concerned, the transresistance pair can be viewed as a degenerate shunt-shunt feedback, and from standard analysis has the following characteristics:

$$\text{Input resistance} = 1/(g_{m2}\, g_{m3}\, R_{o3})$$
$$\text{Output resistance} = 1/g_{m3}$$
$$\text{Transresistance} = 1/g_{m2}$$

where R_{o3} is the aggregate resistance seen at the drain of M_3, being the parallel combination of g_{ds3} and the output resistance of the biasing PMOS cascode current source.

Thus, the overall transfer function of the amplifier can be expressed approximately as:

$$S_{21}(\omega) = \left(\frac{1}{1 + j\omega(C_1 R_1)}\right)g_{m1} \times$$

$$\left(\frac{g_{m3} - j\omega C_2}{(g_{m3} - j\omega C_2)(g_{m2} + j\omega C_2) + (g_{m2} + j\omega C_3)\left(\frac{1}{R_{op}} - j\omega C_4\right)}\right) \times$$

$$0.5\left(\frac{1}{1 + \chi_{buf1}}\right)\left(\frac{1}{1 + \chi_{buf2}}\right)\left(\frac{g_{mbuf1}}{g_{mbuf1} + 0.5j\omega C_{\pi buf2}}\right)$$

where:

$$C_1 = C_{\pi 1} + C_{bondingpad} + C_{\mu 1}$$
$$C_2 = C_{\pi 2} + C_{\mu 3}$$
$$C_3 = C_{PD1} + C_{PD4} + C_{\pi 3} + C_{\mu 1} + C_{\pi 2} + C_{\mu 3}$$
$$C_4 = C_{PD3} + C_{\mu 2} + 0.5C_{\pi,\,buf1} + C_{PD5} + C_{\mu 5} + C_{\mu 3}$$
$$R_{op} = r_{o3} // r_{o5}$$

The χ parameters are the body effect loss parameters for the FET devices, and it should be noted that capacitances C_1 through C_4 are the lumped parasitics at each node in the circuit. C_{PD},S are the parasitic drain/source capacitances, respectively, and the approximations made here are that the multichip module parasitics are small, source followers have operational cutoff near f_t, and the Miller effect on M_1 is negligible (since it sees the low impedance shunt feedback input). This agrees with the low-frequency gain, given by:

$$S_{21,\,LF} = 0.5\left(\frac{g_{m1}}{g_{m2}}\right)\left(\frac{1}{1+\chi_{buf1}}\right)\left(\frac{1}{1+\chi_{buf2}}\right)$$

The -3 dB point can be extracted from this:

$$\omega_{-3dB} = \frac{g_{m3}g_{m2}R_{op}}{(g_{m3}-g_{m2})R_{op}C_2 + g_{m2}R_{op}C_4 + C_{PD1}}$$

which is equal to:

$$\frac{1}{\dfrac{C_4}{g_{m3}} + \dfrac{C_{PD1}}{g_{m3}g_{m2}R_{op}} + \left(\dfrac{1}{g_{m2}} - \dfrac{1}{g_{m3}}\right)C_2}$$

Intuitively, this agrees with the results from zero-valued time constant analysis, with a correction for the pole splitting introduced by C_2 shunted across the gate-drain of M_3. It is important to note that the frequency rolloff is controlled primarily by the parasitic capacitances at the gate of M_3 and at the drain of M_1 (both encapsulated in the C_{PD1} term). M_2 and M_3 are necessarily smaller devices, to minimize the impact of their parasitic capacitance; also, it is desirable that M_2 be small, since the amplifier gain increases with decreasing W in M_2. However, M_1 is necessarily a very large device (400μ/1.2μ), to maximize gain as well as to minimize noise; thus, the parasitic perimeter and area capacitance at its drain determines the overall frequency response of the amplifier. In order to control this capacitance, a ring transistor structure is used in M_1 to achieve a large-W device while reducing drain area capacitance and eliminating drain perimeter cap. For the device sizes shown in Figure 6.7a, a -3 dB cutoff was established at 1.3 GHz. In Figure 6.7b, the simulated and measured frequency response of the amplifier is shown. The upper 3 curves are the gain up through the transresistance stage; the lower one is the aggregate frequency response of the amplifier, including the loss through the output buffers and packaging effects. The 3 curves for just the transresistance stage consist of the SPICE-simulated frequency response assuming an MCM package, the SPICE-simulated frequency response assuming the 52-pin leadless chip carrier package that the amplifier was measured in, and the actual measured data for the gain through the transresistance stage; agreement between the SPICE simulation and the measured amplifier response is quite good. The limitations of the output stages are apparent: an enormous amount of gain loss is incurred through body effect, coupled with the resistive dividers at the input and output (a total of nearly 20 dB of loss). The 1.2 micron process, being limited in both achievable g_m and hence f_t, simply cannot achieve a broadband 50Ω active output match without consuming excessive amounts of power as well as signal loss.

As a broadband amplifier for use in a multichip module, without the limiting output stage, the transconductance-transresistance topology could have achieved its projected +12 dB of gain, and would have been quite usable as the first stage of the receiver chain. However, the multichip module fabrication ultimately proved to be unfeasible, owing to implementation and cost factors. Given the overhead in power and performance required to broadband-match an off-chip, 50 ohm load, a narrowband redesign is called for. The amplifier core is reasonably sound, but a better form of matching – through the use of tuned elements – is needed.

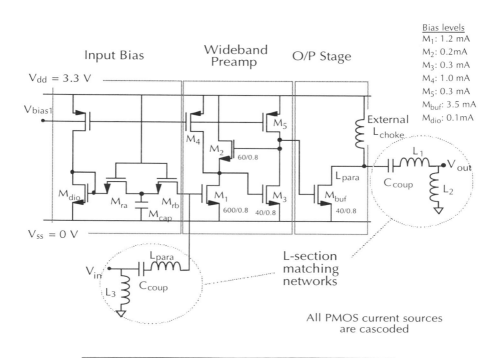

Figure 6.8: Tuned-matching LNA topology

6.3.4. Tuned-Matching Topology

As a second iteration, the output stage of the untuned design is replaced by a common source stage, with an inductor choke plus a tuning network at the output. Simply tuning the output resistance of the broadband stage is hardly optimal; the body effect loss of the output stages will still be present. Instead, a common source output stage is used; in and of itself, it provides another stage of transconductance gain, in addition to the fact that it results in an untuned S_{22} that is nearly equal to unity, and from the maximum unilateral transducer gain equation (Section 6.3.2) this maximizes the overall S_{21} gain of the stage if proper matching is employed. Likewise, input match is similarly implemented; instead of wasting signal power by using a broadband 50 ohm match at the input, proper termination is achieved by matching against the input bias device M_{rb}, which appears to be a passive resistor. The final configuration is shown in Figure 6.8, along with device sizings and bias levels; the matching networks will be described in detail below.

Basically, the input and output matching networks need to match the large input and output resistances to 50 ohms, as shown in Figure 6.8, while resonating the respective input and output capacitances. Interestingly, both input and output are quite similar: a large R shunted with a parallel C, plus a small series parasitic L due to the chip-on-board packaging. The chip-on-board parasitic inductance is extremely small, consisting of a bondwire contributing approximately 2 nH; specific details about the chip-on-board design can be found in the appendix to this chapter. The problem of matching such a network will be approached from two perspectives: use of standard lumped-element analysis, and the use of the Smith chart.

To examine the matching problem from the lumped-element approach, the standard L-section matching network is shown in Figure 6.9. L_1 implicitly lumps the bondwire parasitic, as well as the parasitic inductance of a series AC coupling capacitor. The effective Z_{in} of the network, looking in from the 50 ohm side, is equal to:

$$R_{eq} = \frac{j\omega L_2(j\omega L_1 - \omega^2 L_1 RC + R)}{R(1 + (-\omega^2(L_1 + L_2))C) + j\omega(L_1 + L_2)}$$

The resonance point needs to be set to the desired center frequency; i.e., $1/(\sqrt{C(L_1 + L_2)})$ equated to 1.088 GHz. At this point, R_{eq} is approximately:

$$R\left(\frac{L_2}{L_1 + L_2}\right)^2 = R_{eq}$$

Surprisingly, this behaves quite similarly to a transformer, with turns ratio equal to $L_1/(L_1+L_2)$; in fact, this can be considered to be an inductive transformer. Setting R_{eq} equal to 50Ω yields two equations in the two unknowns L_1 and L_2.

From a Smith chart standpoint, SPICE provides the input/output reflection coefficient plots as shown in Figure 6.10. As before, this is equivalent to a series L, followed by a shunt L to ground. Both the lumped-element method and the Smith chart yield the following values for input and output matching networks, using the 0.8μ extracted layout parameters to determine R and C:

Port	R	C	L_1	L_2
Input	1.8kΩ	3.6pF	5nH	1nH
Output	3.1kΩ	1.1pF	16.6nH	2.4nH

With this in mind, the gain at resonance is approximately given by:

$$
S_{21}(\omega_{res}) = \left(\frac{g_{m3} - j\omega_{res}C_2}{(g_{m3} - j\omega_{res}C_2)(g_{m2} + j\omega_{res}C_2) + (g_{m2} + j\omega_{res}C_3)\left(\frac{1}{R_{op}} - j\omega_{res}C_4\right)} \right) \times
$$

$$
0.5 g_{mbuf} r_{obuf} \left(\frac{L_{2out}}{L_{1out} + L_{2out}} \right) \left(\frac{L_{1in} + L_{2in}}{L_{2in}} \right)
$$

It is similar to that of the broadband case, with the final g_{mbuf} arising from the gain of the output stage. Since the device is tuned, the basic frequency-dependent terms in the equation are from the preamp circuit, which bears the only "untuned" nodes in the circuit.

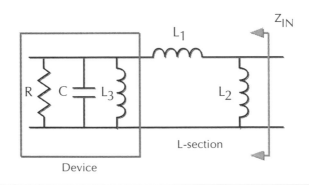

Figure 6.9: Tuned L-section, 50 ohm match
(valid for L_3 an RF choke; R large compared to 50 ohms)

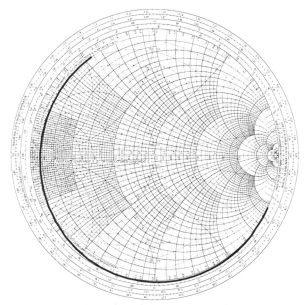

Simulated S_{11}, untuned

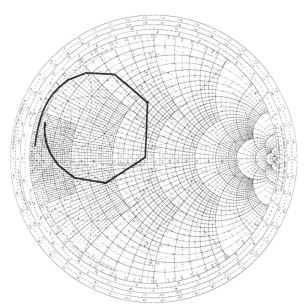

Simulated S_{11}, tuned

Figure 6.10a: SPICE-simulated S_{11} (input) reflection coefficient, tuned and untuned. Plotted from 200 MHz to 1.2 GHz

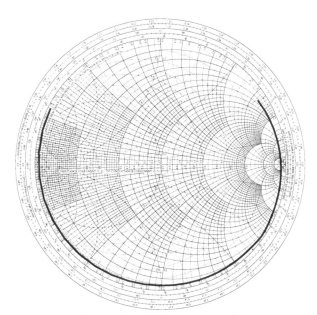

Simulated S_{22}, untuned

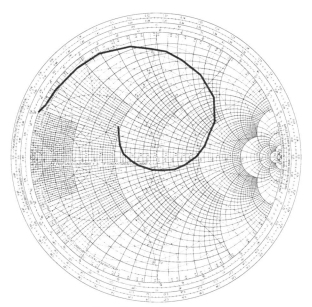

Simulated S_{22}, tuned

Figure 6.10b: SPICE-simulated S_{22} (output) reflection coefficient, tuned and untuned. Plotted from 200 MHz to 1.2 GHz

Lastly, noise performance at resonance must be considered. The input referred noise at the gate of M_1 can be approximated quite well by:

$$v_{ineq}^2 = 4kT\left(\frac{2}{3}\right)\left(\frac{1}{g_{m1}} + \frac{g_{m2}}{g_{m1}^2} + \frac{g_{m4}}{g_{m1}^2}\right)$$

As expected, given the extremely large size of the input device, the noise is dominated by its channel thermal noise; likewise, at these operating frequencies, 1/f noise is not an issue. From this, reflecting back through the input matching network the noise figure of the amplifier can be calculated as:

$$NF(dB) = 10\log\left(2 + 4\left(\frac{L_{2in}}{L_{1in} + L_{2in}}\right)^2 \left(\frac{1}{50\Omega}\right)\left(\frac{2}{3}\right)\left(\frac{1}{g_{m1}} + \frac{g_{m2}}{g_{m1}^2} + \frac{g_{m4}}{g_{m1}^2}\right)\right)$$

6.3.5. Measured Results

The design shown in Figure 6.8 was fabricated and tested as part of the monolithic front-end receiver test chip; instead of the HP 1.2u process, a more advanced 0.8u Hewlett-Packard single-poly, triple-metal digital process became available and was utilized. The measured S_{11} and S_{22} plots are shown in Figure 6.11; they agree well with the simulated S_{11} and S_{22} plots of Figure 6.10. From this, the design was tuned using off-chip L-sections; tuning of S_{11} was highly exacerbated by the need for small, high-tolerance surface mount inductors to tune the large input capacitance, leading to an imperfect match. The measured S_{21} is shown in Figure 6.12, along with the simulated and measured noise figures of the amplifier; the IM3 intercept is -10 dBm for this topology.

After the die was tested, it became apparent that the noise figure was approximately 3 dB high, and the gain was 3 dB low relative to SPICE predictions. The discrepancy was tracked down to the input transconductance device of the LNA; the use of the ring structure resulted in an effective device W that was approximately one-half of what SPICE and hand calculation were utilizing, due to corner fringing effects. Using a corrected W model for the LNA, SPICE simulation agreed well with measured results. Regardless, the agreement gives good indication that the previous analysis is correct, and that RF operation of CMOS circuits is indeed possible.

The conclusion that can be drawn from this is that the preamp structure provides very little in the way of overall gain, relative to that provided by the matching network. In some sense, most of the gain is due to the large input and output resistances; proper tuning allows one to deliver maximum power to these resistors, resulting in maximum voltage gain through the overall amplifier.

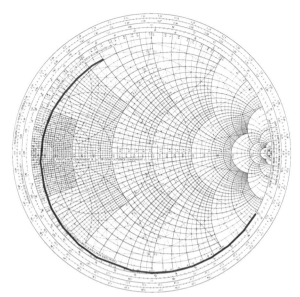

Measured S_{11}, untuned

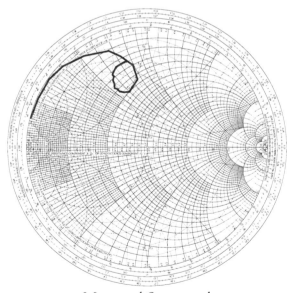

Measured S_{11}, tuned

Figure 6.11a: Measured input S_{11} (input) reflection coefficient, tuned and untuned. Plotted from 200 MHz to 1.2 GHz

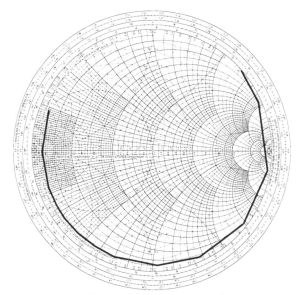

Measured S_{22}, untuned

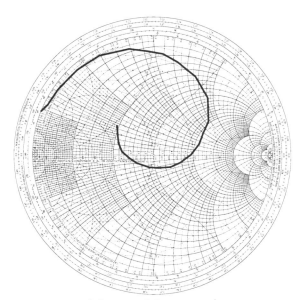

Measured S_{22}, tuned

Figure 6.11b: Measured input S_{22} (output) reflection coefficient, tuned and
untuned. Plotted from 200 MHz to 1.2 GHz

6.4. Sampling Demodulation

Beyond amplification, the second major function required by any receiver front-end is that of frequency conversion of the received RF signal to baseband. When considering demodulation, one very useful method of analysis is to examine the frequency domain representation of the signals; essentially, mixing in the time domain is equivalent to convolution in the frequency domain. In a conventional (superheterodyne) frequency conversion scheme, illustrated by the upper signal path in Figure 6.13, a mixer driven by a local oscillator equal to the carrier ω_c is used to bring the bandlimited signal from RF down to baseband (of course, in a superheterodyne scheme, several stages of mixing may be employed). The resulting baseband signal is then sampled at its Nyquist rate ω_{samp}. Mathematically, if $F(\omega)$ is the original (double-sided) baseband spectrum, bandlimited to $\omega_b = \omega_{samp}/2$, then the RF received signal can be expressed as:

$$\frac{1}{2}(F(\omega - \omega_c) + F(\omega + \omega_c)) \;=\; F_{mod}(\omega)$$

After superheterodyne mixing and Nyquist-rate sampling, the resulting discrete-time sig-

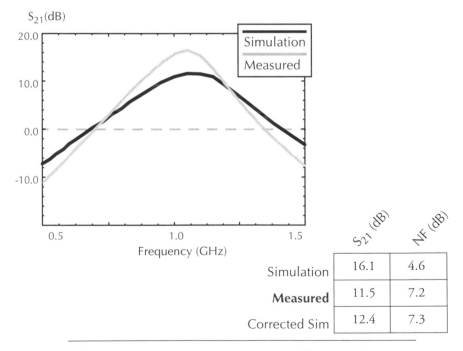

	S_{21} (dB)	NF (dB)
Simulation	16.1	4.6
Measured	11.5	7.2
Corrected Sim	12.4	7.3

Figure 6.12: Measured and simulated amplifier S_{21}, noise figure

nal has aliases at *all* multiples of the sampling frequency ω_{samp}, given by:

$$\left(\frac{2\pi}{\omega_{samp}}\right) \sum_{i=-\infty}^{\infty} F(\omega - i\omega_{samp})$$

However, mixing followed by sampling need not be the only solution for achieving this. Suppose that the received signal $F_{mod}(\omega)$ (eq. 1) is bandpass subsampled at the Nyquist rate ω_{samp} of the baseband signal. After such sampling, the spectrum of the resulting discrete-time signal has a transform equal to:

$$\frac{\pi}{\omega_{samp}} \sum_{i=-\infty}^{\infty} (F(\omega - \omega_{carrier} - i\omega_{samp}) + F(\omega + \omega_{carrier} - i\omega_{samp}))$$

Making the important stipulation that $\omega_{carrier} = k \cdot \omega_{samp}$, where k is an integer, this sum can be simplified to:

$$\left(\frac{2\pi}{\omega_{samp}}\right) \sum_{i=-\infty}^{\infty} F(\omega - i\omega_{samp})$$

which is precisely the same result achieved by the original superheterodyne demodulation structure. The entire demodulation step has been reduced to a single sampling operation, sampled at the Nyquist rate for the *baseband* signal. A graphical depiction of this process is shown in Figure 6.13; the path indicated by the solid black arrows shows the result of each step of a conventional mixer-sampler structure, and the single path indicated by the large gray arrow is the result of bandpass sampling the modulated signal. Clearly, both will yield the same aliased spectrum provided $\omega_{carrier} = k \cdot \omega_{samp}$.

Another interpretation of this result can be seen as follows. The frequency-domain equivalent of sampling is the convolution by a stream of Dirac delta functions, i.e., the effect of sampling the modulated signal with transform $F_{mod}(\omega)$ yields a signal with a Fourier transform equal to:

$$\left(\frac{\omega_{samp}}{2\pi}\right) \int_{-\infty}^{\infty} \left\{ F_{mod}(\omega - \Omega) \sum_{i=-\infty}^{\infty} \delta(\Omega - i\omega_{samp}) \right\} d\Omega$$

Alternatively, this is the same as multiplication in the time domain of the received signal by a sum of equal-power cosines at all harmonics of the sample rate; using the following identity:

$$\sum_{i=-\infty}^{\infty} \delta\left(t - 2\pi\frac{i}{\omega_{samp}}\right) = \left(\frac{\omega_{samp}}{\pi}\right) \sum_{i=0}^{\infty} \cos(i\omega_{samp}t)$$

sampling $F_{mod}(\omega)$ can be equated in the time-domain to:

$$f_{mod}(t) \times \sum_{i=0}^{\infty} \cos(i\omega_{samp}t)$$

Provided proper bandpass anti-alias filtering is done to isolate the modulated signal at $\omega_{carrier} = k \cdot \omega_{samp}$, the required demodulation-and-sampling is thus yielded. Essentially, the mixer has been shown to be redundant, since the sampling operation itself can perform the same mixing function.

Of importance is the fact that the sampling rate does *not* change frequencies, even as the carrier changes. The k in the above stipulation was not specified; so long as the carrier frequency is an integer multiple of the baseband sampling rate, this method works. Hence, a voltage-controlled oscillator is not needed here, and a high accuracy crystal oscillator can

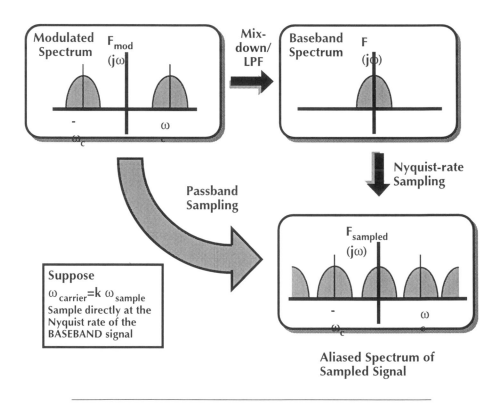

Figure 6.13: Sampling Demodulation

be used instead. The hardware cost is minimal: an accurate switch, implementable using standard MOS technologies, and a fixed-frequency crystal oscillator at the same frequency as the oscillator used to reference the transmitter's synthesizer PLL. In terms of power, the demodulator is passive, dissipating energy only in the crystal oscillator and the resistive channel of the MOS device, and is certainly much lower than an equivalent full mixer/oscillator configuration.

Theoretically, the sampling demodulator is an excellent implementation for the demodulator in terms of hardware and power, utilizing an active device as a simple passive switch. However, before any judgment of real usefulness is passed, its practicality for use in a transceiver for personal communications must be assessed: it must be able to demodulate a signal centered at ω_c=1.088 GHz, while sampling in excess of ω_{samp}=128 MHz for timing recovery. The above analysis was performed symbolically; the specific choice of carrier frequency and sampling rate will be examined below.

6.4.1. Quadrature Recovery and the Choice of Carrier Frequency

Since the sampling introduces mixing with equal-power cosines, the in-phase signal is easily recovered, as in the above derivation. However, the quadrature phase requires a sine wave in the mixing operation to successfully recover the information. Clearly, some more work is needed than just the simple sampling step, since the quadrature data will be lost. One immediate method is to utilize an additional sampling operation, shifted in phase by $\pi/2$. By doing so, the time-domain representation of this additional sampling operation can be expressed as:

$$f_{received}(t) \times \left\{ \left\{ \left[\sum_{i=0}^{\infty} (-1)^i \cos(2i\omega_{samp}t) \right] + \left[\sum_{i=0}^{\infty} (-1)^i \sin((2i+1)\omega_{samp}t) \right] \right\} \right\}$$

The proper mixing is yielded if the condition is made that $\omega_{carrier} = k \cdot \omega_{samp}$, and k is odd. Otherwise, the second sampling step simply yields the in-phase signal again. Thus, use of such a system with quadrature modulation requires a second sampling unit offset by a simple 90° phase shift.

If the parameters of the system are considered - a 64 Mchip/sec DQPSK spread-spectrum signal, requiring a minimum sampling rate of 128 MHz to perform timing recovery - the above stipulation that k be odd places some severe restrictions on the choice of carrier frequency. In particular, near 1 GHz the only possibilities are 896 MHz and 1.152 GHz. Eliminating this restriction is critical, since neither of these frequencies is usable for the system. Instead, a carrier of 1.088 GHz is chosen, with a pair of interleaved 128 MHz baseband sampling switches. Since interleaving is performed, there is a 180° phase relationship between the two switches, instead of 90° as described above. Given that a noninteger frequency relationship is established between the carrier and the baseband sampling rate, the impact of this on the demodulation process must be considered.

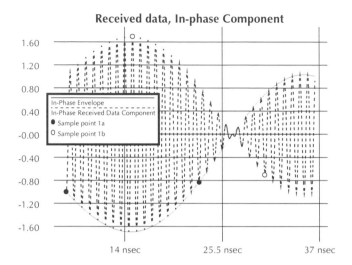

(a) In-phase received data component;
I-sampling demodulation points are marked at 1/128 MHz

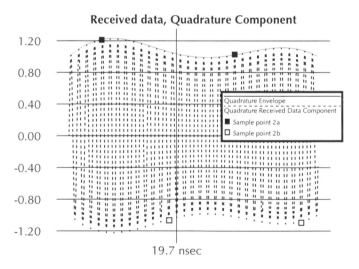

(b) Quadrature received data component;
Q-sampling demodulation points are marked at 1/128 MHz

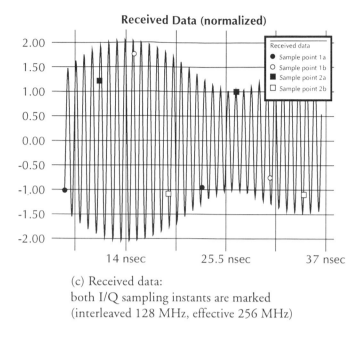

(c) Received data:
both I/Q sampling instants are marked
(interleaved 128 MHz, effective 256 MHz)

Figure 6.14c: Time-domain view of sampling demodulation. Note the recovery
of both in-phase and quadrature data components, and the timing relationships
between samples 1a and 2a. Figure (c) shows the complete received signal, as well
as both sets of sampling points.

Because of the interleaving, the aggregate effective sampling rate of the pair is 256 MHz;
relative to 1.088 GHz, the demodulated output should have a frequency offset equal to
1.088 - 1.024 GHz = 64 MHz. As with the premodulation of the FIR filter in Chapter 4,
a 64 MHz cosine wave sampled at 256 MHz yields a sequence {... +1, 0, -1, 0, ... },
whereas a 64 MHz sine wave sampled at 256 MHz yields a sequence { ... 0, +1, 0, -1, ... }.
Thus, the interleaving is correct: due to the zero samples, one sampler output will have
entirely in-phase information, with alternating sign bits due to the +/-1, and the other
will have entire quadrature information. In Figure 6.14, the demodulation process is illustrated graphically. The received modulated in-phase and quadrature data streams are
shown, with the dotted lines indicate the envelope of the baseband data and the sample
points marked for each demodulator. Clearly, one sampler recovers the in-phase data,
whereas the other recovers the quadrature. The sign alternation, or demodulation from
64 MHz down to baseband, can be removed digitally by simply toggling the sign bit after
analog-to-digital conversion

There are two important points that need to be made:

- Since the final 64 MHz demodulation is done in the digital domain, there is clearly no DC information content in the analog signal. Circuit impairments such as 1/f noise and DC offsets in the analog circuitry can be filtered out without affecting the recovered signal.

- The demodulator will have a pair of interleaved sampling switches, one for in-phase and one for quadrature. However, their behaviour is symmetric: there is nothing intrinsic in the architecture to determine that one sampling path is "in-phase" or one sampling path is "quadrature". Such assignments can only be made after timing recovery has been performed in the digital CDMA receiver. This fact is utilized heavily in the design of the digital CDMA receiver timing recovery loop: in particular, no timing-recovery feedback is necessary between the digital and analog sections of the receiver, resulting in several significant hardware reductions (Chapter 7).

Lastly, some intuition should be forthcoming, given the elegance of the result. In actuality, 1.088 GHz does maintain an odd-integer, 90° phase relationship with the baseband signal: it is an odd multiple of 64 MHz. If samples 1a and 2a of Figure 6.14 are considered exclusively as the output, it is apparent that these will simply be the original data signal, in-phase and quadrature. The addition of output samples 1b and 2b provides in-phase and quadrature data outputs at another sampling phase for timing recovery; because of the carrier modulation, those other samples will have an additional 180° phase shift relative to samples 1a and 2a, requiring a simple digital sign toggle to correct.

6.4.2. Bandpass Antialiasing

Furthermore, careful consideration needs to be given to the image rejection filtering at RF. Since the received signal is being subsampled, this image rejection filtering now performs an antialiasing function above and beyond the typical anti-image function normally found in RF transceiver systems. The need for this is clear: the demodulator will perform frequency conversion at any integer multiple of ω_{sample}, not just the desired one. The 128 MHz oversampled interleaving described in the previous section has a second benefit: as in baseband analog systems, oversampling decreases the antialiasing requirements in the system: the nearest frequency images that need to be filtered from the system are a 960 MHz and 1.216 GHz, as opposed to 1.024 GHz and 1.152 MHz. Likewise, the broadband thermal noise introduced by the LNA will be spectrally folded if it is not rejected before the sampling switch. As shown in Figure 6.2, two filter stages are used, on either side of the LNA. Not only do two filtering stages increase rejection, but the post-LNA placement of the second filter rejects the out-of-band thermal noise introduced by the LNA itself. Unfortunately, since neither multichip module packaging nor on-chip inductors were available, the second filter stage must be off-chip, nominally matched to a 50 ohm impedance level from the LNA. Given that a typical RF ceramic filter response [Murata93] has the characteristics as shown in Figure 6.15, allowing an extra 64 MHz in the antialias rolloff translates to another 30 dB (for two stages) in adjacent channel rejec-

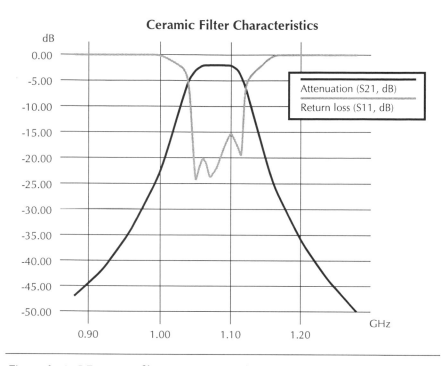

Figure 6.15: RF ceramic filter attenuation and return loss characteristics [Murata93]

tion at the first alias frequency, critical if the receiver is to coexist with high-power cellular phones in the 800 MHz band, pagers in the 930 MHz band, and cordless telephones between 902 and 928 MHz.

Lastly, the overall system design provides additional rejection filtering against interference from aliased images. As described in Chapter 2, the intrinsic property of spread-spectrum is that of rejecting in-band additive noise: exactly the problem posed by spectral aliasing. With respect to the signal post-correlation, the effective impact of the aliased images has been lessened by the processing gain. In other words, the filter skirts in Figure 6.15 are effectively 18 dB lower than what would be found in a non-spread-spectrum system employing sampling demodulation. The ultimate choice of using spread-spectrum in this system was hinged upon the additional filtering gain provided with respect to aliasing in the sampling demodulator: another example of the impact that coupling analog, digital, and system design had on the overall wireless system architecture.

6.4.3. Frequency Stability and Incoherent Demodulation

To ease the hardware requirements, the necessity of zero phase and frequency offsets must be carefully considered. If possible, carrier recovery should be avoided at all costs, which leads to the requirement that very accurate frequency references are needed in both the transmitter and receiver, and immunity to nonzero phase offsets.

First, the assumption that $\omega_{carrier} = k \cdot \omega_{samp}$ in the transmitter must be examined. Surprisingly, this is not a difficult requirement for the hardware; a PLL synthesizer automatically achieves this condition, and does so with extreme accuracy when referenced to a crystal oscillator. Second, although the accuracy of crystals is extremely good (typically 1 part in 10^5), it is not perfect: the transmitter and receiver will be frequency (and phase) offset by some amount. From the data sheets on the crystals to be used in this system [RFM90], worst-case frequency mismatch between transmitter and receiver is specified as 25 kHz; from physical measurements on pairs of crystals, this is in actuality quite pessimistic: in the sample set available, the worst-case measured offset was actually 3 kHz. Using the 25 kHz value, for a baud rate of 1 MHz this translates to a worst-case phase error of 9° from baud-to-baud, assuming no carrier recovery in the receiver - a small error, considering the impact of intercell and intracell interference in the overall system.

The simplest solution to this problem is to utilize a phase-differential QAM scheme, which encodes the transmitted information on the transition between constellation points instead of the constellation points themselves - hence the choice of DQPSK for the system modulation format. For voiceband systems, carrier recovery is absolutely necessary: 25 kHz is an enormous offset relative to a 32 kHz channel. For high-speed data communications, 25 kHz is minuscule compared to the 1 Mbaud channel signalling rate, and solutions such as differential phase encoding are viable, leading to both power and complexity minimization. This is an example of how a digital encoding at baseband allows the elimination of costly analog hardware; removal of absolute phase dependence is achieved with only a minor expense in digital processing, and no expense in bandwidth.

Lastly, the phase noise of the crystal oscillator itself must be considered. Such jitter can be modeled as a phase error in the recovered/sampled output signal, given that it arises from random fluctuations in both the transmitter and receiver crystal output frequencies. As a first-order analysis, supposing that the signal to be sampled is given by $A_v \cdot \cos(2\pi \cdot 1.024 \cdot 10^9 t)$, sampling at 128 MHz with zero phase difference should yield a constant (DC) output. To compute the impact of phase noise on the recovered signal, the worst-case will be at the point of maximum slew, or at the zero crossing of the sinusoid (Figure 6.16). Theoretically, the output should be constant zero; instead, it yields some random output noise signal, whose amplitude can be approximated as being linearly related to the jitter, resulting in an SNR due to jitter noise equal to $20 \log(2/\omega_c t_{jitter})$.

Thus, a worst-case estimate on the allowable jitter window can be realized. For a sampling jitter SNR of 30 dB[1], at 1.024 GHz with a 128 MHz sampling rate, this translates to an allowable jitter of +/- 28 psec about the true sampling point, or approximately

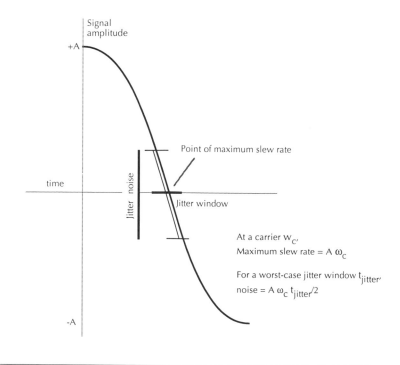

Figure 6.16: Translation of phase noise into additive signal noise in the demodulation process.

1 part in 300. Although it seems like a simple matter for a crystal oscillator to achieve this, the design challenge arises in maintaining this accuracy through the necessary on-chip clock buffering, especially in the presence of supply bounce, substrate injection, and capacitive coupling effects.

---

1. The linearity of the A/D converter is only required to be 4 bits from system simulations, due to the spread-spectrum nature of the signal; see Chapter 7. The sampling jitter SNR is chosen to be 6 dB beyond this, hence the 30 dB requirement.

6.5. MOS Sampling Demodulator Design

Of course, the sampling described in the above theory is an idealization; no device can ever achieve "delta-function" precision in sampling, since it takes a finite amount of time for the sampling switch to close. In this section, the practical aspects of designing such a demodulator will be addressed, from the perspective of switch track bandwidth, noise performance, and distortion. The sample-and-hold (S/H) itself must have be able to track input waveforms with a bandwidth exceeding the carrier frequency, which lies between 1 and 2 GHz. For a silicon MOS sample-and-hold, there are several significant non-idealities that must be considered. In Figure 6.17, a basic MOS S/H is displayed, along with a plot of a signal to be sampled and the output of the S/H. Essentially, the MOS device is a simple on-off switch, passing charge to the sampling capacitor when in track mode, and preserving the charge when in hold mode.

First, when the S/H is in track mode, the switch is closed and the voltage across the sampling capacitor should ideally equal the input voltage. However, the drain-source resistance R_{ds} of the MOS device is nonzero, equal to:

$$R_{ds} = \frac{1}{\mu C_{ox}\left(\dfrac{W}{L}\right)(V_{gs} - V_t)}$$

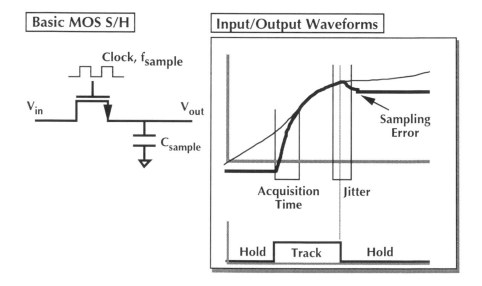

Figure 6.17: MOS Sample-and-Hold Characteristics

where the device parameters are the same as described in the discussion of f_t above. The effect of R_{ds} is to create an R-C lowpass filter with C_{sample}, hence limiting the bandwidth that the system can track accurately. Likewise, the acquisition time t_{acq} needed to switch from hold to track mode is determined by this lowpass filter, and thus $t_{acq} = (1/\text{Track BW}) = R_{ds} \cdot C$. For a sampling demodulator, this must be well in excess of the desired carrier bandwidth, requiring that both R_{ds} and C be significantly smaller than what would normally be done in a sampling array. Both of these implications are rather dire. A small R_{ds} implies that the sampling switch needs to be large for a fixed supply voltage, exacerbating the charge injection and feedthrough effects described below. What is more important is that a small C results in a proportionally larger kT/C noise, impacting the effective noise figure of the demodulator. Both of these will be discussed in detail with regard to the final design of the sampling demodulator.

Second, as the switch moves from track mode to hold mode at the sampling instant, two mechanisms enter to degrade the signal: feedthrough from the sampling clock, and charge injection from the switch itself. Feedthrough from the sampling clock is a result of the gate overlap capacitance C_{ovl} and the sampling capacitor forming a divider; if the sampling clock edge is falling from a supply voltage V_{dd} to ground, this results in a sampling error equal to:

$$\Delta V_{feedthrough} = -\left(\frac{C_{overlap}}{C_{sample}}\right)V_{dd}$$

Likewise, charge stored in the channel of the MOS switch must be removed, and hence is injected directly onto the sampling capacitor. Assuming that the charge divides evenly between the source and drain of the switch, which is reasonable given the fast switching edge necessary for demodulation, this results in a sampling error ΔV equal to:

$$\Delta V_{charge} = \frac{-Q_{chan}}{2C_{sample}} = \frac{-WLC_{ox}(V_{dd} - V_{in} - V_t)}{2C_{sample}}$$

It is important to note that both of these effects are inversely proportional to C_{sample}, and proportional to W. Given that W needs to be large for a small R_{ds}, and C_{sample} needs to be small to achieve the track bandwidth, the system must be able to tolerate a large ΔV at the sampling instant.

Of course, the above discussion is highly dependent on the fabrication technology. Ignoring the constant sampling error due to clock feedthrough for the moment, the above analysis yields the classical figure-of-merit for a MOS device [Gray89], known as the speed-accuracy product, equal to $\Delta V_{charge} \cdot t_{acq}$, and can be simplified to $2L^2/\mu$. Since minimizing this is the key goal, and (like the device f_t) is proportional to the square of the channel length, as devices scale the performance of MOS devices improves dramatically. For 4-bit accuracy on a peak-to-peak voltage swing of 1 V, a track bandwidth of 1.8 GHz, and $\mu = 200$ V/(cm^2-sec), a minimum drawn L of 0.25 microns is required.

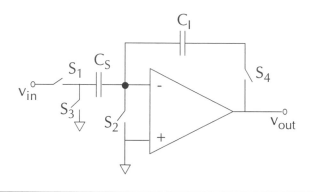

Figure 6.18: Bottomplate sampling topology (single-ended)

Obviously, a 0.25 micron process is not readily available at this time, and would thus seem to preclude the use of direct subsampling as a means of RF demodulation from a carrier of 1 GHz. However, the above result is quite pessimistic: it assumes that the charge injection pedestal ΔV onto the capacitor is entirely signal dependent, with no simple means of elimination. An improved sampling switch topology is shown in Figure 6.18; a technique known as bottom-plate sampling is employed to remove the signal dependence in the error term. The operation can be understood as follows: At first, S_1 and S_2 are closed, allowing the input signal V_{in} to be sampled onto the capacitor C_S. When S_1 and S_2 are opened, an amount of charge proportional to the signal voltage and to the size of capacitor C_S is left floating on C_S exactly as in the previous circuit. However, if S_2 is opened slightly before S_1, then the S_2 switch determines the sampling instant. Since both the drain and source of transistor S_2 are at ground, no signal dependent charge injection is introduced. When S1 is subsequently turned off, the charge in the channel sees an open circuit on the other side of C_S and therefore all the signal-dependent charge that is injected must flow to the other side of the transistor to the input source (and again no error is introduced). Thus, the injection pedestal can be made constant; by employing differential signalling, this constant offset simply becomes a harmless common-mode shift. Differential signaling also eliminates the constant offset due to clock feedthrough. In effect, the performance limits of such a differential bottomplate sampling topology are established by second-order effects, most notably differential mismatch [Lin90]. The switch accuracy does remain quadratically dependent on L: however, it is vastly improved over the simple topology of Figure 6.17 by the differential bottom-plate nature of the switch. Given careful layout, a sampling demodulator operating at 128 MHz, demodulating a 1.088 GHz signal, has been demonstrated with a 0.8 micron CMOS technology available today.

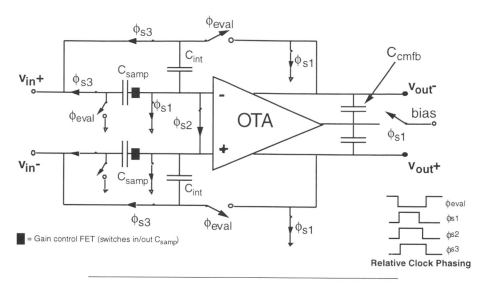

Figure 6.19: Variable-gain, bottomplate sampling
demodulator topology

6.5.1. The Bottomplate Sampling Demodulator

The sampling demodulator that was designed and fabricated is shown in Figure 6.19.
Also shown is the operational transconductance amplifier (OTA), required to effect charge
transfer from the sampling capacitors and buffering into the next stage. Furthermore, the
sampling capacitor has been divided into two separate capacitors, C_{int} and C_{samp}, with
C_{int} serving a dual role as the integrating capacitor for the OTA, and C_{samp} can be
switched in and out of the switch array by FET switches. This "splitting" of the sampling
capacitor allows a variable-gain function to be achieved by the sampling switch; with C_S
present, the conversion gain of the mixer is +3 dB, without C_S, the conversion gain is
0 dB. Since the issues of variable-gain amplification, frequency response, and settling
characteristics of the OTA are really baseband phenomena, discussion of the OTA design is
relegated to the next chapter; the design of the switch array is the principle focus of the
following discussion, with regard to track bandwidth, mixer noise figure, and linearity.

To determine the track-mode bandwidth of the switch, the track mode configuration of
the sampling demodulator is shown in Figure 6.20. Both capacitors are assumed to be
present, giving the worst-case bandwidth condition due to the additional parasitics involved, and the equivalent linearized model is shown in Figure 6.21. From extracted device parameters, the -3 dB bandwidth (voltage transfer onto the capacitors) is 2.1 GHz,

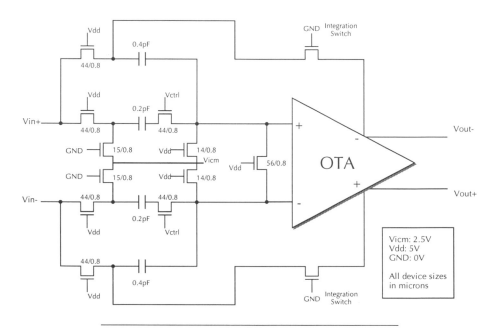

Figure 6.20: Sampling demodulator,
track-mode configuration.

both by hand calculations and SPICE simulation: the overdesigned track bandwidth is to
account for extraction errors (especially with regard to analog parameters such as g_m), as
well as provide some margin for phase balance between the each leg of the differential
switch array. The penalty paid for the lack of a second polysilicon layer is evident here;
the capacitor is formed by a metal1-metal2-metal3 stacked structure (Figure 6.22), with
the parasitic capacitor associated with the metal1-metal3 top-plate being approximately
half of the main capacitance. As a result, the design value for C must be significantly
smaller than what would be possible in a double-poly process, impacting the kT/C performance dramatically; the design values for the integrating and variable-gain sampling
capacitors are 0.4 and 0.2 pF respectively.

The noise figure of the mixer is thus straightforward to determine: the noise mechanism
is simply the well-known kT/C noise of the switch itself [Gray89]. Given that the capacitors need to be extremely small to admit the 1 GHz received signal, it is not surprising
that the noise figure of the mixer is poor. For the design values of $C_S + C_I = 0.6$ pF[1], this
results in an equivalent total mean-square voltage noise equal to $6.9 \bullet 10^{-9} \overline{v^2}$. Conserva-

---

1. This is assuming the weakest signal situation (maximum gain), where both C_S and C_I will be
   employed in the demodulator.

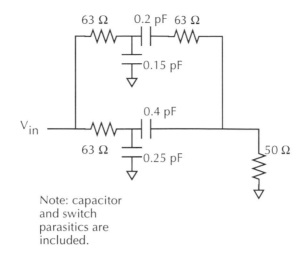

Note: capacitor
and switch
parasitics are
included.

Figure 6.21: Track-mode linearized model of the sampling
array (equivalent half-circuit)

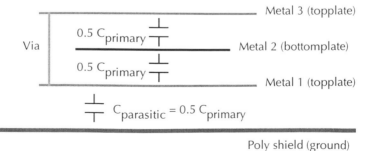

Figure 6.22: On-chip capacitor structure

HP 0.8u non-double-poly process:
Area cap (M1-sub,M1-M2, M2-M3): 0.035 fF/sq.micron;
Perim cap (all metal layers): 0.049 fF/micron.

tively assuming a total signal bandwidth of 128 MHz (to account for filter rolloff), the kT/C noise can be referenced to the total in-band noise of a 50Ω resistor, leading to an equivalent noise figure of 18.1 dB. To compensate for this, the architecture of the receiver in Figure 6.2 places an additional 16 dB low-noise buffer stage just before the demodulator. The extreme noise figure resulting from kT/C noise is perhaps the key disadvantage of employing a direct-sampling scheme for RF downconversion - the total broadband noise in the switch is admitted into the system. If some means were available to filter out the switch thermal noise, potentially by using an on-chip inductor to realize a bandpass network, the noise figure of the mixer could be reduced tremendously. Alternatively, with CMOS technology scaling, the R_{ds} of the switch could be made commensurately smaller, allowing larger C's to be used in the sampling operation.

Lastly, distortion in the sampling demodulator needs to be considered. Surprisingly, the demodulator is quite linear; the linearized model of Figure 6.21 applies over a large input range. Intuitively, the demodulator reduces to a FET switch and a capacitor; a high track bandwidth implies that the operating frequency is such that very little voltage is lost across the nonlinear resistance of the switch. Thus, until the input signal is large enough that the switches themselves begin to turn off, the demodulator should behave in a extremely linear fashion. The capacitor of Figure 6.22 is a strictly linear device; the only nonlinearity is that of the sampling switches. The above analysis agrees with simulation (Figure 6.23): the simulated third-order distortion intercept of the modulator is 20.4 dB.

6.5.2. On-Chip Clock Buffering and Layout Issues

Lastly, the issue of how the sampling clock is generated must be addressed, with respect to the 28 psec jitter accuracy needed in the sampling operation. As described previously, this requirement can easily be met by the off-chip crystal oscillator; however, preserving this phase-noise performance on-chip is critical. In particular, proper buffering, coupled with careful layout considerations, is required to insure that the clock edges remain as jitter-free as possible, as well as minimizing substrate injection from the clock edges.

First, for system-level considerations, a 128 MHz off-chip clock oscillator is chosen; sampling is done on both rising and falling edges, with the in-phase demodulator clocked on the rising edge and the quadrature demodulator clocked on the falling edge. Furthermore, the off-chip oscillator signal is chosen to be a differential sinusoid instead of a square-wave pulse. The differential nature of the signal minimizes radiation and coupling at the board level; the sinusoidal nature minimizes the harmonic energy at 1 GHz, important given that a high-sensitivity LNA is on the same die. Of course, this signal must be converted to a square wave before it can be utilized on-chip, requiring a differential analog input clock buffer. The clock buffer for the chip is shown in Figure 6.24; it is embedded as part of the clock input pads. The center pad is used to bias the differential buffer; the nominal external bias for this node is 1 mA, for operation at 128 MHz. Likewise, the three sampling edges required by the differential bottomplate sample/hold must be synthesized from the converted clock signal, by the NAND-based nonoverlapping generator circuit shown in Figure 6.25. Again, symmetric layout is employed, with signals switching differentially between the phi1 and phi2 clock phases.

In addition to the buffering, the layout of the sampling switches themselves is crucial. Figure 6.26 shows the die fabricated to test the sampling demodulator by itself, along with several important features of the layout. First, the clock lines are run down the centerline of the design, with the in-phase and quadrature demodulator blocks on either side. This digital "no-man's land" is heavily shielded, with grounded metal-3 and poly shields above and below to minimize coupling through the air, as well as coupling into the substrate by the clock lines. Second, to guarantee the high-speed falling edge on the sampling switches, very large local clock buffers are placed next to each switch array; the sizing of the final inverter is a $78\mu/1\mu$ PMOS, $30\mu/1\mu$ NMOS. A simulated 10-90 fall time of 110 psec (40 psec to switch turn-off) is produced by these local buffers, necessary to accurately sample the gigahertz-band input signal.

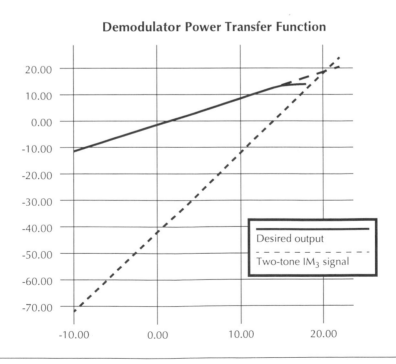

Figure 6.23: Simulated distortion characteristics of the sampling demodulator. Two equal-power input tones were used for the IM_3 test.

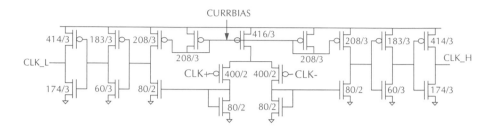

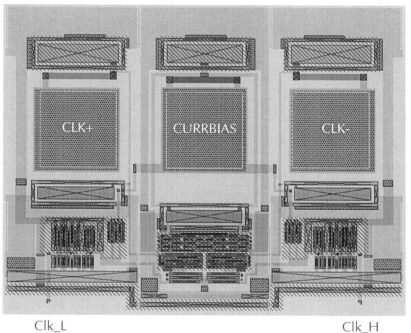

Figure 6.24: Schematic and layout of differential input
clock buffer. Device sizes are in lambda
(0.5 micron = 1 lambda)

Furthermore, large grounded guard rings are placed around all of the fast digital switching devices, as to capture substrate injection as early as possible. Although ground-lead
inductance going off-chip typically limits the effectiveness of the grounded guard rings,
all effort was made to minimize this inductance, most notably by the presence of a back-
side substrate ground contact and direct vias to the ground plane in the chip-on-board
packaging (see the appendix to this chapter). Lastly, large on-chip MOSFET-gate bypass capacitors (100 pF) are placed near each demodulator, to minimize the impact of supply
bounce on the jitter performance of the clock buffers.

6.6. From RF to Baseband

In this chapter, the RF characteristics of a custom, integrated CMOS analog receiver front-
end, suitable for use in the mobile end of the broadband CDMA downlink. Beyond achieving functionality normally relegated to silicon BJT's or GaAs FET's, such as low-noise amplification, the cornerstone of this new integrated architecture is the use of sampling
demodulation to achieve frequency conversion, where the harmonics of a high-speed,
low-jitter baseband clock edge are used to mix down the received signal. Although such a
demodulator may appear to be a simple CMOS differential bottomplate sampling switch,
the particular characteristics required to sample a 1 GHz modulated signal adds several
layers of complexity and consideration to issues such as track bandwidth, charge injection, and switch linearity.

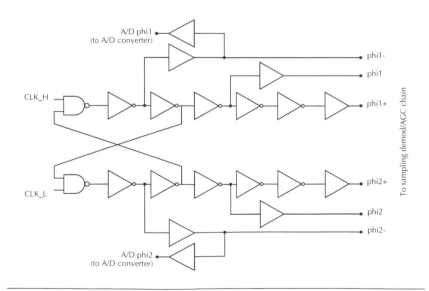

Figure 6.25: On-chip nonoverlapping multiphase clock generator
(3 edges, both rising and falling phases)

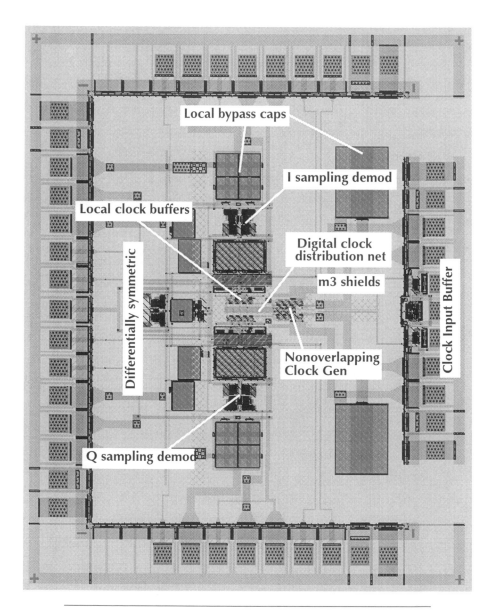

Figure 6.26: Standalone sampling demodulator testchip, illustrating several points on die layout.

With the opening of the sampling switch, the signal has traversed the boundary from the continuous-time, gigahertz-band domain, into the discrete-time, baseband domain, where switched-capacitor techniques are dominant. The next chapter describes the second half of the analog front-end: the analog hardware required to perform baseband variable-gain amplification, and A/D conversion, completing the list of functions that need to be performed in the analog domain. Although the results for the prototype LNA have been discussed in this chapter, the testchip results for the complete demodulator will be deferred until the end of Chapter 7, where the entire analog (RF and baseband) receive chain can be examined.

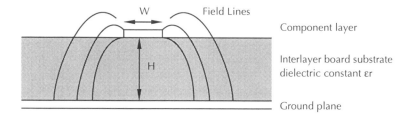

Figure 6.27: Board transmission line structure

6.7. Appendix: RF Chip-on-Board Design

This appendix is intended to be an overview of the board design process, especially in consideration of gigahertz-band operation and the need for chip-on-board packaging for parasitic control. It is not intended as a tutorial; it is assumed that the reader has some passing knowledge of electromagnetism and board design.

6.7.1. Electromagnetic Effects

To provide a good 50 Ω matched conductor on the board, microstrip and stripline structures must be employed. Creation of microstrips is relatively simple in a multilayer board; if a ground plane is placed beneath an RF trace, a microstrip is naturally created, as shown in Figure 6.27. Since the electromagnetic fields are half in air and half in the board material, an effective dielectric constant $\varepsilon_{r,eff}$ is calculated with respect to the substrate ε_r:

$$\varepsilon_{r,\ eff} = \frac{\varepsilon_r + 1}{2} + \frac{\varepsilon_r - 1}{2} \frac{1}{\sqrt{1 + 12\frac{H}{W}}}$$

Since wave propagation in microstrips can be approximated as quasi-TEM, an effective inductance and capacitance per unit length can be assigned, with an equivalent characteristic impedance Z_o equal to:

$$Z_o = \frac{60}{\sqrt{\varepsilon_{r,\ eff}}} \ln\left(\frac{8H}{W} + \frac{W}{4H}\right)$$

The key thing to note is that Z_0 is effectively controlled by the W/H ratio of the stripline; the larger the W/H, the lower the impedance. Furthermore, the smaller the ε_r, the higher the impedance; alternatively, for a smaller ε_r a larger W/H ratio will be needed to keep the same characteristic impedance Z_0.

Figure 6.28 lists typical materials that are used in board fabrication. Although it ha significantly inferior loss tangent and higher ε_r, FR4 is the material of choice: Duroid and cyanate ester are very brittle, having little water content, and it is difficult to make boards of any significant size. Also, at least for the boards that have been designed to test the custom integrated circuits, the traces are fairly short (around 5 cm maximum length), and there are only a few RF traces (since the circuits are highly integrated). Thus, signal loss due to skin effect and nonzero board substrate loss tangent is negligible; from a practical standpoint, multilayer FR4 is used for all of the board designs.

As far as choice of Z_0, 50Ω is usually used for the following reasons:

- Standard coax cabling is readily available at 50Ω

- All available test equipment will terminate and match in 50Ω

- 50Ω is realizable on the board: to achieve 50Ω in FR4, a W of 39 milliinches (mils) over an H of 19 mils (W/H ~ 2.0) is needed. This allows tolerance in the board fabrication process, which is typically +/- 1 mil on H and +/- 0.5 mil on W. For example, to achieve $Z_0=200\ \Omega$, W/H must equal 0.027; for W=10 mils, an H of 370 mils is needed, far too thick to be practical, as well as resulting in parasitic higher-order transmission modes along the line.

Material	ε_r	Loss tangent/ dissipation factor
FR4	4.5	0.019
Duroid	2.20	0.0009
63N cyanate ester	3.3	0.009

Figure 6.28: Available board fabrication materials

6.7.2. General Layout Issues

A few rules work well:

- Keep RF traces as short as possible, and minimize the number of corners on RF traces; each corner will radiate. If trace widths need to change, tapering of the trace is effective in preventing reflections on the line; this is known as a tapered transformer [Pozar90].

- Use supply planes whenever possible. For example, the RF test board described below uses 3 supply planes: V_{dd} (analog), V_{dd} (digital), V_{icm} (common mode bias voltage). Not only does this minimize supply bounce/ripple by providing a physically short, low-impedance return path, but also simplifies routing on the board tremendously. Separate digital and analog supplies to minimize coupling, either by splitting the planes, or by using separate planes.

- The order of the board layers is important: make certain that the second layer of the board is a ground plane, otherwise the microstrip structure does not exist. Even if the second layer were a supply plane, supply planes tend to have more noise than ground planes, due to nonzero source resistance and inductance. Shield the RF traces from any digital switching lines by the digital lines on the solder side (bottom) of the board. By placing several supply planes between the critical analog traces and the digital output lines, better shielding is achieved (Figure 6.33).

- Route the RF traces by hand before any sort of automated routing. This is certainly CAD-tool dependent, but as a general rule hand routing will result in shorter RF traces and fewer corners.

- Use coaxial connectors to bring the supply voltage on-board. RF signals tend to couple into everything unless shielded; a long unshielded wire coming from the bench power supply is a perfect antenna, and can result in excessive RF noise coming from the supply.

- Capacitor bypass DC supply and control pins as much as possible. As discussed below, the choice of capacitor size varies critically with the application.

6.7.3. Discrete Inductors and Capacitors

Due to high-frequency effects, special consideration must be given to passive inductors and capacitors for use in RF boards. Not only are such devices needed for supply bypassing and AC coupling purposes, but they also become necessary for filter design: at 1 GHz, the wavelength (30 cm free-space) is still large, making striplines unattractive as a means of filtering while maintaining a small form factor. Fortunately, at 1 GHz, discrete passives can still be used; parasitic effects have not become dominant. The key issue is the self-resonance frequency of the devices: capacitors form a series tank circuit with the bondwire inductance of the package, whereas surface-mount spiral inductors have an intrinsic self-capacitance from winding turn to winding turn (Figure 6.29).

L (nH)	Nominal Q @ 500 MHz	Series R (ohms)	Self-Res. Freq (GHz)
4.7	20	1	2.0
6.8	20	1	2.0
10.0	20	1	1.0
15.0	20	2	1.0
22.0	20	2	1.0
33.0	10	5	0.850

Figure 6.30: Table of some commercially-available surface-mount inductors

Figure 6.30 lists several commercially available surface-mount inductors, along with their self-resonant frequencies (SRF). For operation at 1 GHz, an SRF in excess of 2 GHz is needed, which limits the usable inductance value for designs to around 10 nH. Another issue is that of inductor tolerance: small values of L are difficult to achieve, owing to the parasitic inductance introduced by the solder ball on the surface mount board itself (on the order of 0.25 - 0.5 nH of inductance). Due to this, and the limiting parasitics in-

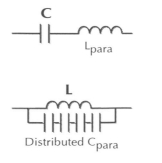

Figure 6.29: Parasitics in surface-mount discrete capacitors, inductors

volved with getting a signal on-and-off of a silicon chip, on-chip inductors are highly desirable. Unfortunately, on-chip inductors were not realizable in the available silicon fabrication technology, and thus significant effort was expended in "hand-tweaking" the board to achieve the accurate inductances.

An even more critical issue is the choice of capacitors, since they provide both AC signal coupling and DC bypass for the supply and control lines. The usual "low-frequency" design mentality is to use as large of a capacitor as possible for both functions; at RF, this is not viable owing to the limitations placed by the SRF. Shown in Figure 6.31 is a plot of measured SRF as a function of capacitance value, for three families of capacitors (1206, 0804, 0603 size form factors). The dependence illustrated here is interesting; since SRF = $1 / \sqrt{L_{para} C}$, one would expect a square-root dependence on C and similarly dependent on the package size (controlling L_{para}). From the measured plot, size has little impact, whereas the SRF is inverse-linearly dependent on the capacitance value. The reason for this lies in the fabrication of the surface mount capacitor itself: to achieve high values of C, many layers are wound within the package itself, thus leading to larger parasitic L's. Thus, the parasitic inductance is strongly controlled by the capacitance value itself, not by the physical size of the package.

Thus, using a large capacitor is hardly optimal, since at 1 GHz the capacitor will appear to be strongly inductive. However, for both AC coupling and DC high-frequency bypassing, the capacitor needs to appear as a zero impedance. Since AC coupling applications only require a zero impedance at the operating frequency, the parasitic series resonance can fortuitously be used to achieve this, since it results in a near-zero impedance at the SRF, as shown in Figure 6.31b. Furthermore, the Q is fairly low; SRF tolerance is not critical; thus, by choosing capacitors with an SRF near the operating frequency of 1 GHz, a good AC couple can be achieved. For the RF testboard described below, all AC coupling capacitors were 0603 form factor, 10 pF capacitors. DC bypass applications, on the other hand, are more difficult. Although the SRF-technique will provide a low impedance for a range of frequencies at RF, bypassing at lower frequencies is still necessary. Thus, two or more capacitors are usually used in parallel as shown in Figure 6.32, with one capacitor providing low-frequency bypass and the second providing RF bypass. However, care must be taken: there is a parallel LC formed by the parasitic L of C_1 and the actual capacitance of C_2. This implies that, at some resonant frequency, the bypass network itself is a nearly open circuit, and will be subject to ringing at that frequency. Damping of this parallel LC with a shunt R is necessary to prevent this.

6.7.4. Chip-On-Board Design

For large pin count die, the packaging itself limits circuit performance tremendously: package inductance, plus large lead inductance, sets fairly low upper bounds on the maximum operating frequency, especially when the circuit input capacitance is large.

- High integration implies high lead count, and hence large packages. Since the entire analog section was to be integrated into a single die, the RF sections plus the baseband analog processing required a 132-pin ceramic quad flat pack to

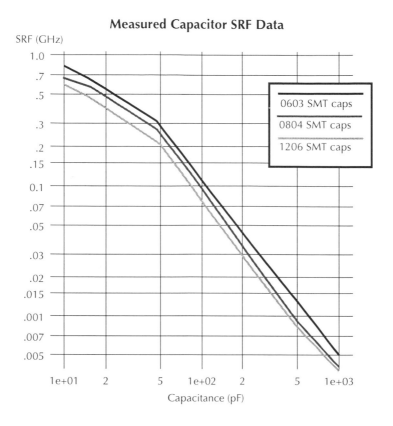

Measured Capacitor SRF Data

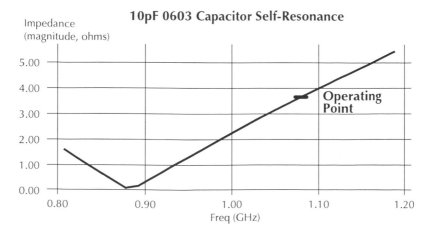

Figure 6.31: Impact of inductive parasitics on discrete capacitors at RF.

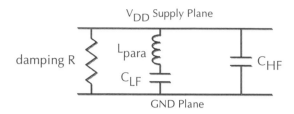

Figure 6.32: Board-level bypass strategy. Typical values are:
C_{LF} = 0.1 µF, C_{HF} = 10 pF, R=500Ω

accommodate all of the necessary signal, control, and supply/ground lines. Such packages were measured to have a worst-case package + bondwire inductance of 20 nH, and a lead frame capacitance of approximately 2 pF.

- For CMOS, a large input capacitance is nearly unavoidable, especially in light of the 0.8u process available: the LNA input device, a 600u/0.8u NMOS, exhibits a C_{gs} of 1.3 pF, while the bonding pad results in another 0.5 pF of additional input capacitance.

The result of this is that the packaged version of the design limits the LNA maximum operating frequency to approximately 600 MHz (the self-resonance point of the effective series L-C). Furthermore, supply and ground lead inductance effects are similar; substrate injection and on-chip coupling cannot be minimized at 1 GHz, since the board ground plane is separated from chip ground by the package inductance.

Obviously, a better packaging technology is needed. Multichip modules are certainly one solution; however, a viable manufacturing process could not be obtained. "Good" high frequency packages have 50 ohm controlled microstrips for the leads, as well as local supply bypass capacitors on the physical package itself; unfortunately, these packages are prohibitively expensive. The final solution is particularly elegant: eliminate the package [Lau94]. Direct die-attach to the surface mount board is quite possible, easily integrated into the regular surface-mount process, and inexpensive. Details of the chip-on-board process used will be described below.

6.7.5. RF Testboard

The following documents the surface mount board that was developed for testing the entire monolithic analog front-end, as described at the end of Chapter 6. The physical fabrication specifications of the testboard are as listed in Figure 6.33. In Figure 6.34a, the board layout is shown, with several salient features marked. A magnified view of the bonding site for the analog front-end testchip is shown in Figure 6.34b; for bondability, a

1) BOARD FILM SET DEFINITION

Layer1 | Component Signal, SEE BELOW

Layer2 | Power Plane GND, 1oz CU

Layer3 | Power Plane 1 (analog VDD), 1oz CU

Layer4 | Power Plane 2 (analog VICM), 1oz CU

Layer5 | Power Plane 3 (digital VDD), 1oz CU

Layer6 | Signal 4 (digital signals only), 1oz CU

Layer7 | Ground Plane (digital GND), 1oz CU

Layer8 | Solder Side (digital signals), 1oz CU

SMC | Solder Mask, Component Side

SMS | Solder Mask, Solder Side

SILKC | Silk Screen, Component Side

----

2) SIZE: 5500 mils x 4700 mils

3) MINIMUM LINE/WIDTH SPACING: 3 mils

4) Laminate is FR-4 (Green)

5) Overall board thickness is 0.075 +/- 0.008

6) Comments/Special Handling

This is a chip-on-board design; since the die needs to be attached and bonded directly to the component layer, the comp layer needs to be plated with wirebondable gold on the component layer, 50 microinches thick with 150 microinches of nickel, 0.25oz foil

Also, layers 2, 3, 4, 5, and 7 are gnd/power planes; hence, the gerbers are negative images. Interlayer spacing between component layer and powerplane GND should be 21 mils, all other spacings should be minimum.

All dimensions are in English units unless otherwise specified

Figure 6.33: Board fabrication parameters
(as submitted to the fabrication facility)

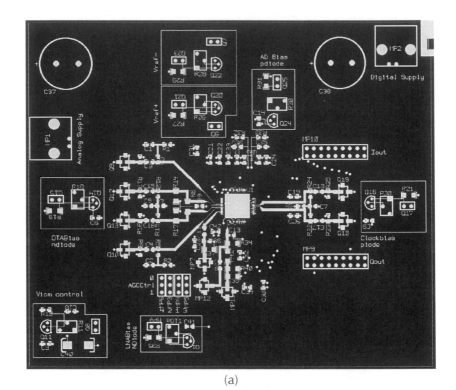

(a)

(b)

Figure 6.34: RF testboard layout. (a) Layout;
(b) close-up of the chip-on-board bonding area.

(Test jig for the analog front-end prototype die)

4 mil[1] wide line is used for the board bonding lands, with the minimum allowable 3 mil space between them. Likewise, these board bonding lands are pitch-matched one-to-one against the pads on the die. This makes bonding particularly simple, as well as making certain that every pad has the same (minimum) bondwire length and associated parasitic inductance. With a 7 mil pitch, this determined the pad pitch on the die of 150 microns. Furthermore, it is not sufficient just to be able to bond to the board; the signal must be able to connect to other components on the board. Thus, an interleaved set of vias (Figure 6.35) is incorporated on the board to provide connection points. In particular, these vias can directly contact the ground and supply planes, providing very high-performance, low inductance (< 1 nH) connections. The geometry indicated in Figure 6.35 represent the minimum spacings allowable by our board fabrication technology.

As far as the actual die attach site goes, it comprises of a single square plane of gold-copper metal; with a die size of 280 mils by 280 mils; the actual bonding site has a 20 mil overhang on all sides to allow some tolerance in die attach (attach area is 320 mil by 320 mil). Furthermore, a direct ground via is connected at one corner of the bonding site; this provides a solid ground directly underneath the die. By using conductive epoxy to attach the die, a good back-side contact can be formed between the ground plane and the silicon die substrate. This is critical in minimizing the effects of on-chip substrate injection, since a path to a solid low-impedance ground is provided through the back-side. The fabricated board photo is shown in Figure 6.36.

---

1. From contact with various board fabrication/assembly houses, 4 mils is the minimum width that can be reliably bonded with technology available at this time.

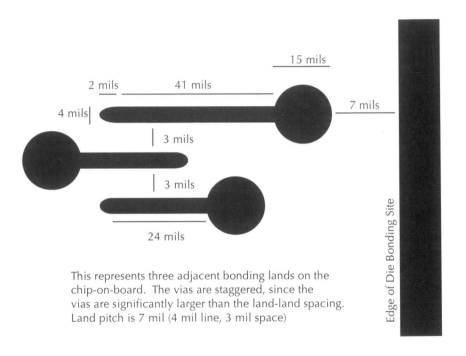

This represents three adjacent bonding lands on the
chip-on-board. The vias are staggered, since the
vias are significantly larger than the land-land spacing.
Land pitch is 7 mil (4 mil line, 3 mil space)

Figure 6.35: Interleaved via layout geometry of the chip-on-
board bonding lands.

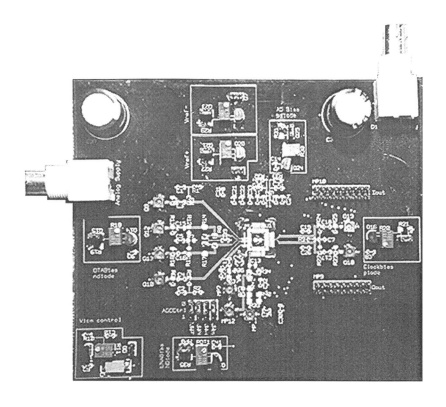

Blowup of
chip-on-board
die attach region

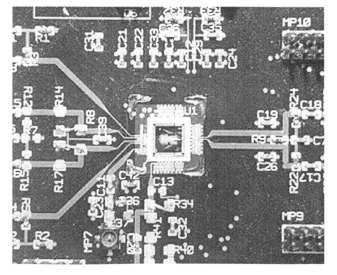

Figure 6.36: Fabricated board photograph

REFERENCES

[Bakog90]   H.B. Bakoglu. *Circuits, Interconnections, and Packaging for VLSI.* New York: Addison-Wesley Publishing Co., 1990.

[Behzad95]   A. Behzad. *The Implementation of A High Speed Experimental Transceiver Module with an Emphasis on CDMA Applications.* M.S. Thesis, U.C. Berkeley, Berkeley, CA, 1995.

[Gonz84]   G. Gonzalez. *Microwave Transistor Amplifiers: Analysis and Design.* Englewood Cliffs, NJ: Prentice-Hall, 1984.

[Gray84]   P.R. Gray and R.G. Meyer. *Analysis and Design of Analog Integrated Circuits, 2nd Edition.* New York: John Wiley and Sons, 1984.

[Gray89]   P.R. Gray. Course Notes for EECS 290Y, Analog-Digital Converter Technologies. Department of EECS, U.C. Berkeley, Berkeley, CA. Spring 1989.

[LNA81]   H. Fukui, ed. *Low-Noise Microwave Transistors and Amplifiers.* New York: IEEE Press, 1981.

[Lau94]   J. Lau. *Chip On Board: Technologies for Multichip Modules.* New York: Van Nostrand-Reinhold, 1994.

[Lin90]   Y-M. Lin. *Performance Limitations of High-Resolution Video-Rate Analog-Digital Interfaces.* Ph.D. Dissertation, Memorandum No. UCB/ERL M90/55, U.C. Berkeley, Berkeley, CA, June 19,1990.

[MCM91]   R. Johnson, R. Teng, and J. Balde, eds. *Multichip Modules: System Advantages, Major Constructions, and Materials Technologies.* New York: IEEE Press, 1991.

[Murata93]   Murata-ERiE Corporation. *Microwave Ceramic Filter Product Guide, Catalog No. M-10-B.* Smyrna, GA, 1993.

[Pozar90]   D.M. Pozar. *Microwave Engineering.* New York: Addison-Wesley Publishing Co., 1990.

[RFM90]   F. Perkins, R.F. Monolithics. *High-Speed Video Dot Clocks for High-Resolution Computer Display Applications, Application Note.* Dallas, TX, 1990.

[Viter95]   A.J. Viterbi. *CDMA: Principles of Spread Spectrum Communications.* Reading, MA: Addison-Wesley Pub. Co., 1995.

[Watts89]   R.K. Watts. *Submicron Integrated Circuits.* New York: John Wiley and Sons, 1989.

7 The Receiver: Baseband Analog Processing

with Lapoe Lynn

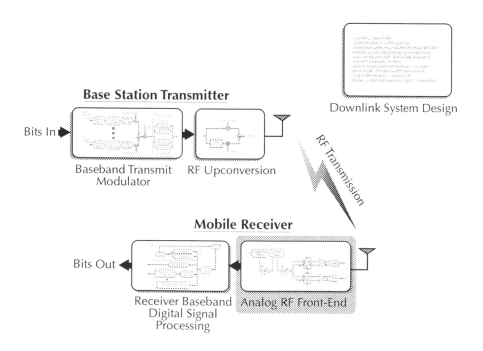

Downlink System Design

With RF amplification and demodulation performed, the remaining functionality of the receiver front-end consists of automatic gain control, and analog-to-digital conversion. As discussed in Chapter 6, the ability to leverage off complex digital processing results in an architecture that favors moving the analog-to-digital (A/D) converter as early in the receive signal path as possible: in effect, creating an "A-to-D front-end". Of course, this implies that an extremely high-speed, low-power A/D converter is necessary. One of the key elements of this chapter is the design of a 4-bit, 128 MHz A/D converter which consumes only 2.2 mW, implemented in a 0.8 micron standard single-poly CMOS process. At first glance, a resolution of 4 bits seems quite low; it will be shown that, at the system level, the quantization noise associated with a 4-bit converter can indeed be tolerated, given the spread-spectrum nature of the signal. This is crucial to low-power design: minimizing converter resolution is tantamount to minimizing its power consumption, and every consideration is made at the system level to keep the required number of bits down to the absolute lowest possible value.

However, this implies that automatic gain control (AGC) and variable-gain amplification (VGA) are required in the analog domain. The explanation for these requirements is simple: the mobility of the terminal within a fading environment implies large and rapidly changing shifts in the received signal amplitude. A moving terminal may be physically very close to the transmitter (and therefore receiving a strong signal), and then move very far away (where it receives a weak signal); it may also suddenly move into a fading null, or a location where there is no line-of-sight path between the base station transmitter and the mobile. Thus, the AGC serves to guarantee that the signal presented to the A/D converter circuitry appears to be of constant power regardless of the actual signal strength at the antenna. Thus, the two issues of converter resolution and dynamic range are decoupled: the system can tolerate fairly low resolution, but requires considerable dynamic range due to fading. While it is true that bits in the A/D may be traded off for gain in the AGC (i.e., increased resolution in the A/D, beyond the minimum required for SNR, allows it to detect a smaller signal), higher resolution converters become exponentially costly in terms of power, area and complexity, especially at the required sampling speeds. Given the constraint of low-power consumption, keeping the A/D converter at its lowest possible resolution and adding variable-gain functionality is the correct design tradeoff. Thus, the other key element of this chapter is the implementation of the variable-gain circuitry required by the front-end. Every AGC contains two critical blocks – the VGA, and the power detector circuit that feeds back the control signal used to adjust the gain to the correct level. At the time of this writing, the power detection block for the AGC loop has not been implemented; however, the VGA is far more difficult from an implementation standpoint, given that the power detector only needs to track the variation in the gain (hundreds of Hertz). The VGA needs to pass the full bandwidth of the received signal, while maintaining low power consumption.

Thus, to complement the discussion of the RF circuitry in Chapter 6, this chapter is devoted to discussion of the design and implementation of the baseband variable-gain amplification and analog-to-digital conversion circuitry required in the analog front-end; more detailed documentation of the circuits discussed below can be found in [Lynn95].

7.1. System Dynamic Range and Linearity Requirements

Of course, the immediate question that arises is, "How many bits?" To answer this, system-level simulations of the overall receiver were performed, with respect to the impact on quantization noise on the ultimate system BER. Shown in Figure 7.1 is the BER curve as a function of users, under statistically severe fading conditions and varying bit-quantization in the A/D converter. Clearly, no improvement in the curve is seen for greater than

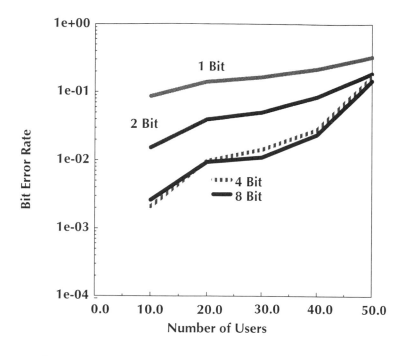

Figure 7.1: BER vs. number of in-cell users, as a function of A/D bit quantization (extreme fade condition, no error correction)

4 bits of resolution: given that the desired user's signal is already "buried" in the multipath noise of other users, beyond 4 bits the quantization noise ceases to be significant. Thus, a baseband linearity requirement of 4 bits is established for the entire discussion of this chapter; as will be seen, even 4 bits of linearity is quite difficult to achieve given the available fabrication technology.

However, it seems counterintuitive that a system that can intrinsically accommodate 64 users only requires 4 bits of resolution in the A/D converter itself for proper operation. The answer lies in the processing gain provided by the spread-spectrum decorrelation: in some sense, high-speed A/D conversion is traded against resolution requirements. More specifically, the 18 dB of processing gain provided by the spreading translates into another 3 bits of effective resolution at the output of the spread-spectrum correlators in the digital back-end. Ultimately, with respect 1 MHz symbol decisions being performed by the receiver, a 7-bit dynamic range converter is being employed.

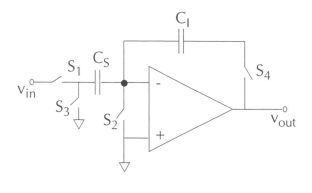

Figure 7.2: Bottomplate sampling topology (single-ended)

Lastly, the overall system dynamic range, from channel measurements and simulations discussed in Chapter 2, is 45 dB. Part of the VGA functionality is embedded in the RF elements of the front-end: the LNA itself can be turned off or on, providing an extremely coarse two-setting (+16 dB or -2 dB) form of gain control. Likewise, a second "off-on" RF buffer stage is provided at the input to the sampling demodulator, again providing a coarse form (+17 dB or -2 dB) of gain control. The remaining 12 dB, however, must be much more precise than this: the variable gain topology described below is to implement this fine-grain control, being digitally controllable from 0 to 12 dB in 3 dB steps.

7.2. Variable-Gain Amplifier Topology

As mentioned before, the design of the receiver VGA is a significant departure from traditional designs. Because of the inherent sampling incorporated into the mixing operation, a multi-stage cascade of discrete-time amplifiers is used to replace a standard continuous-time VGA. Therefore, what was once an amplifier design problem involving a trade-off between gain and bandwidth becomes a sample-and-hold design requiring an op-amp and a set of switched capacitors that can settle to four bits of accuracy within one clock period. As described in Chapter 6, Figure 7.2 shows a basic bottomplate sample-and-hold amplifier: however, such a simplistic design is inadequate for practical use. First, the design must be made differential. A differential circuit is necessary not only to reject the charge injection from the switches, but also to reject common-mode noise coupling into the analog signal path. Figure 7.3 shows how digital switching can cause common mode ringing on analog lines. If the sample is taken single-ended, a large error would result – an error often larger than the signal itself. However, a differential design with careful layout, with balanced parasitic capacitances, can maintain a small differential signal amidst such a large common mode interfering signal.

When performing bottom plate sampling, the operational amplifier shown in Figure 7.2 is used to buffer and drive the output to the correct value, proportional to the input voltage. However, settling is a critical issue: when the circuit is in track mode, the output no longer follows the input voltage. Therefore, when the hold phase arrives, the output must settle to the correct output voltage from some reset value, nominally zero. If the amplifier is appropriately designed with adequate phase margin, the settling curve follows an exponential curve with a single RC time constant. Typically, only half of the clock cycle time is dedicated to the hold period (the other half goes to the tracking period). Therefore, the output has half of one cycle (T/2) to exponentially approach its final value. Unfortunately, an infinite amount of time is required for an exponential curve to reach its final exact value, given by $V_{out} = A(1 - e^{-t/\tau})$, where A is the final output value and τ is the time constant of the circuit. Therefore, there will always be an error introduced into the signal due to incomplete settling. Fortunately, the situation is not quite as grim as it may seem. The settling error, $Ae^{-t/\tau}$, is linearly proportional to the final value A if t is constant. In other words, if given the same amount of time to settle, the circuit will always settle to the same percentage of its final value. Therefore, the gain of the circuit will always be reduced by the constant factor $e^{-t/\tau}$. As long as t is constant, this gain compression does not introduce a real error into the signal. The actual exact value of the gain is rather unimportant since the negative feedback from the AGC control loop will try to force the output of the VGA to be constant regardless of the gain of each stage, although the compression still causes a loss of gain in the signal path. However, a real error does enter the signal path when the allowed time, t, is not constant. Timing jitter in the sampling instant provided by the edge of a clock causes t to vary somewhat from sample to sample, resulting in an error that is proportional to the jitter variation, D_t, and to the slope of the settling curve at the sampling instant. Given enough time or a very fast circuit, this error can be quite negligible (since the slope of the curve decreases with time). Unfortunately, the receiver design has neither of these luxuries. The 128 Msample per second requirement translates

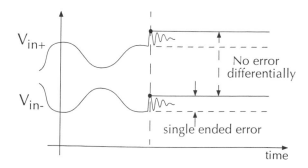

Figure 7.3: Coupling of digital ringing into differential signal paths.

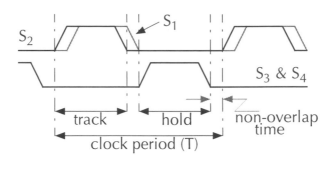

Figure 7.4: Non-overlapping clock phases used to
control sample & hold.

into a 7.8 nanosecond clock cycle time. At most half of this (and in reality, much less than half) can be used for settling time. Figure 7.4 shows the clock phases for the circuit in Figure 7.2. The hold period (and therefore, the settling time) is reduced in length by the non-overlapping period between phases (necessary for proper sampling) as well as by the finite rise and fall time of the clock's edges. The hold period for the receiver's clocks is about 2.5 ns. In order to both avoid degrading the SNR of the signal, as well as to avoid significant gain attenuation, a the settling error in the receiver VGA should be kept smaller than half the LSB of the 4 bit converter. In other words, the error must be less than 1 part in 32 (3.125%) of the signal level:

$$e^{-t/\tau} = 0.03125$$

$$t/\tau = \ln\left(\frac{1}{0.03125}\right) \cong 3.47$$

The above show that settling to this level of accuracy requires approximately 3.5 settling time constants. With t= 2.5 ns, this gives τ on the order of 0.7 ns. However, since the VGA design consists of more than one stage, each settling error will add to the others, resulting in a larger error. For example, in order to keep the error lower than one-half LSB for a four stage VGA, each stage would need to settle to less than 1 part in 128 (~0.8%) of the exact value. The result is a τ on the order of 0.5 ns, or over 310 MHz of bandwidth.

Therefore, a careful design of the amplifier is required with settling time (and therefore, closed loop bandwidth) as a primary design constraint. One of the main limiting factors in the design of a high speed sample-and-hold amplifier (SHA) is the operational amplifier used to drive the output to its final value. In order to achieve the fastest possible circuit, a single-stage transconductance topology has been chosen for the op-amp. The details of

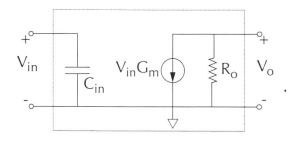

Figure 7.5: Two-port representation of the
operational transconductance amplifier

the design of this operational transconductance amplifier (OTA) are described later in this
chapter; however, for the purposes of this section, it is sufficient that the OTA can be modeled as a G_m transconductance with a high output impedance R_o, as shown in Figure 7.5.

Figure 7.6 shows the SHA during the evaluation phase of its operation. C_I encloses the
OTA in a series-shunt feedback loop with feedback factor:

$$f = \frac{C_I}{C_I + C_S + C_p + C_{in}}$$

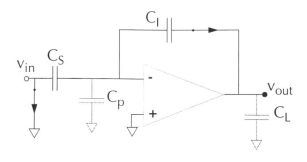

Figure 7.6: Amplifier in hold mode (evaluation phase).

where C_p is the parasitic capacitance at the summing node, C_{in} is the input capacitance of the op-amp, and C_L is the output load being driven. Given that the open loop gain of the amplifier is $G_m R_o$, the closed loop gain and frequency response of the circuit can be determined. Including the capacitive divider before the input to the OTA, the total closed loop gain can be expressed as:

$$A_{CL} = \left(\frac{C_S}{C_S + C_P + C_{in} + C_I} \right)\left(\frac{A_{OL}}{1 + A_{OL}f} \right) = \left(\frac{C_S}{C_S + C_P + C_{in} + C_I} \right)\left(\frac{G_m R_o f}{1 + G_m R_o f} \right)\frac{1}{f}$$

$G_m R_o f$ is the loop gain T of the circuit and f is the feedback factor as above. Substitution yields the expression:

$$A_{CL} = \left(\frac{C_S}{C_S + C_P + C_{in} + C_I} \right)\left(\frac{C_I + C_S + C_P + C_{in}}{C_I} \right)\left(\frac{T}{1 + T} \right) = \frac{C_S}{C_I}\left(\frac{T}{1 + T} \right)$$

The loop gain term in the above equations is close to unity for large values of T. However, in an AGC, a small reduction in the gain is quite irrelevant since the overall gain in the receiver is affected very little by this variation because of the negative feedback of the control loop. Therefore, lower values of T are acceptable, and the closed loop gain of the amplifier in is close to (but not exactly) C_S / C_I. The settling time at the output node is determined by the output impedance of the closed-loop amplifier and by the output load. Therefore, the RC time constant at the output of the amplifier is given by:

$$\tau = \left(\frac{R_o}{G_m R_o f} \right)C_L = \frac{C_L}{G_m f}$$

Clearly, in order to maximize the speed of the amplifier, the feedback factor, f, should be made as large as possible, and the transconductance, G_m, should also be increased if possible. G_m is an op-amp parameter, and its optimization is discussed below. However, the feedback factor relates directly to C_S and C_I. In order to achieve a gain greater than one, the closed loop gain of this topology, C_S / C_I requires that C_S be larger than C_I, reducing the feedback factor by the same increase in gain. An alternative topology is shown in Figure 7.7 in which the integrating capacitor is not only used to close the feedback loop around the OTA, but is also used as a second sampling capacitor to capture a sample of charge from the input. During the track mode (as shown in Figure 7.7) the two capacitors are shorted together in parallel, effectively making one large sampling capacitor. When the hold, or integration phase arrives, the switches shown in the figure are reversed and the charge on C_S is transferred to C_I. The closed loop gain of the circuit is now $(C_S + C_I / C_I)$. In other words, the new topology achieves a larger loop gain for the same feedback factor and bandwidth (a gain of one has been added to the old C_S / C_I gain term). Alternatively, the sampling capacitor can be made smaller for the same closed loop gain, but higher bandwidth. (e.g. C_S must equal C_I to effect a gain of two in the new topology, but in the previous design C_S had to be twice as large as C_I – decreasing the feedback factor.) The final design of the SHA for the VGA is shown in Figure 7.8. The center sampling switch across the op-amp's inputs and an extra clock phase, ϕ_{s2}, have been added to provide better charge injection matching from the sampling operation (charge injection from the center switch will be less dependent on geometry matching than charge from two separate switches). Also shown in Figure 7.8 is the control for changing the gain of this stage. When the indicated switch is closed, then the circuit is essentially the same as the circuit in Figure 7.7 whose operation was just described as having a closed-loop gain of (C_S +C_I / C_I). However, when the switch is opened, C_S is removed from the circuit. The closed-loop gain of the circuit now becomes C_I / C_I. Therefore, each stage of the final VGA consists of a SHA providing a gain of either (C_S +C_I / C_I), or a gain of unity. This capacitive ratio can be chosen to be any value desired; however, increasing C_S also decreases the feedback factor and therefore the bandwidth. Due to the extremely high speed requirements, stages implementing only 3 dB each have been cascaded together to provide 12 dB of total gain. Three dB of gain (or approximately 1.4x) requires a C_S equal to slightly less than half of C_I.

Finally, the sample-and-hold design, being fully differential, requires common mode feedback to ensure that the common-mode voltage at the output remains fixed. Figure 7.8 shows the capacitive common-mode feedback that has been added with a switch for reset during the sampling phase. This feedback is essentially the same technique used when the C_I loop is closed (except, of course that it is common mode), since the capacitors feed back the common mode output to the tail current source of the OTA, as will be seen in the next section.

7.3. Operational Amplifier Design

The heart of the sample-and-hold amplifiers described in the last section is the operational amplifier. Clearly, the high-speed requirements of the system dictate a simple, high-bandwidth design for the OTA. While most CMOS sample-and-hold amplifiers (for use in a pipeline A/D for example) use two-stage amplifiers to achieve higher gain, the receiver VGA design does not have this requirement. Interstage gain amplifiers in a pipeline A/D require gain accuracy on the order of the resolution of the entire converter (in other

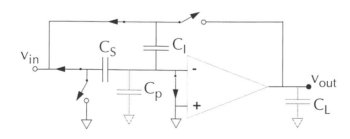

Figure 7.7: Sample-and-hold topology for sampling onto C_I as well
as C_S.

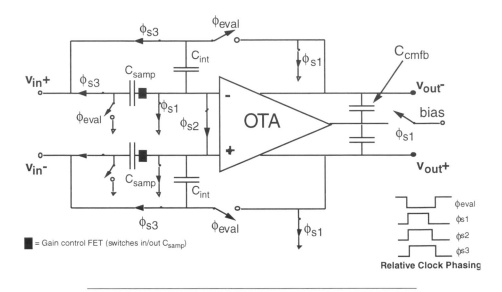

Figure 7.8: Final sample-and-hold topology

words, each stage of an N-bit pipeline A/D must have gain precisely controlled to one part in 2^N) [Conroy94]. However, the negative feedback provided by the overall control loop compensates for any error in the gain of individual stages of the AGC. Therefore, in the trade-off between gain and bandwidth, a lower-gain-but-higher-bandwidth design has been chosen. The effects of gain compression from finite OTA gain are illustrated in Figure 7.9 for a pipeline stage and in Figure 7.10 for the receiver VGA.

A single-stage telescopic cascode topology is the fastest known op-amp topology available in a CMOS technology. The use of common gate cascode transistors eliminates the Miller effect at the inputs, and the low impedance seen looking into the source of a cascode transistor means the circuit is essentially a single-pole system. While a folded cascode has the advantage of increased headroom capability, the inclusion of PMOS transistors in a folded design adversely affects the non-dominant poles. This lower PMOS f_T limits the bandwidth when feedback is applied. Therefore, the telescopic cascode in Figure 7.11 has been designed for use as an operational transconductance amplifier [Uehara93]. The input transistors and cascode transistors (M_{in+}, M_{in-}, M_{ncasc+}, and $M_{n-casc-}$) are n-type devices to maximize the bandwidth of the amplifier. The width of the input devices is determined by a settling time optimization analysis presented later in this chapter. The PMOS transistors M_{psrc+}, M_{psrc-}, M_{pcasc+}, and M_{pcasc-} form a pair of active current source loads in order to achieve a high output impedance and therefore a high DC gain. The size of these devices is determined primarily by the required signal swing at the output of the OTA. The devices down the middle of the diagram in Figure 7.11 represent a high-swing

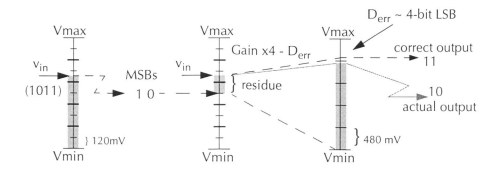

Figure 7.9: 2 bit-2 bit pipeline with small interstage gain error introduced

bias circuit for the two NMOS cascode transistors. M_3 and M_4 are simply a cascoded PMOS current mirror to bias the two NMOS transistors. M_1 is forced to operate in the triode or linear range of operation by the diode connected transistor M_2. Therefore, M_1 acts as a source degeneration for M_2, and the W/L ratio of M_1 is chosen so that the voltage drop from drain to source will match the desired V_{ds} across the two input devices. This value, V_{ds_input} should be as low as possible without pushing the input devices into the triode region of operation. Therefore, the final bias voltage at the gates of the NMOS cascode devices is designed so that the input transistors M_{in+} and M_{in-} have drain to source voltage: $V_{ds\_input} = V_{ds\_1} = V_{dsat} + V_{margin}$. A very aggressive margin of 150 mV has been chosen

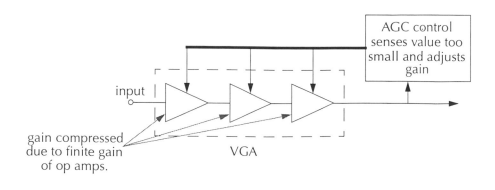

Figure 7.10: AGC compensation of gain errors in the feedforward receiver path.

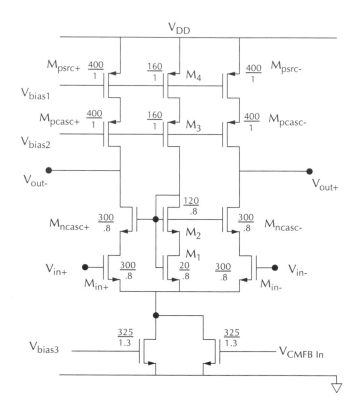

Figure 7.11: Telescopic cascode amplifier. All device sizes are in microns.

so that the available headroom for the signal to swing in the negative direction is maximized as long as M_2 is designed properly to match the V_{gs} drop of the cascode transistors. Therefore, the design equations for M_1 and M_2 are:

$$W / L_2 = \left(\frac{W / L_4}{W / L_{psrc}} \right) (W / L_{ncasc})$$

$$I_{D1} = \mu C_{ox} (W / L_1) \left((V_{gs1} - V_T) V_{ds1} - \frac{1}{2} V_{ds1}^2 \right)$$

The above can be solved for W/L_1 since I_{D1}, $V_{gs} - V_T$, and the desired V_{ds1} are known. The bias circuitry for the PMOS current sources have been designed with the same technique, as shown in Figure 7.12. Figure 7.12 also shows the tail current source bias generated from an off-chip current source flowing into an NMOS diode. This diode serves as a reference for all currents used by the OTA's. The tail current source of the amplifier simply

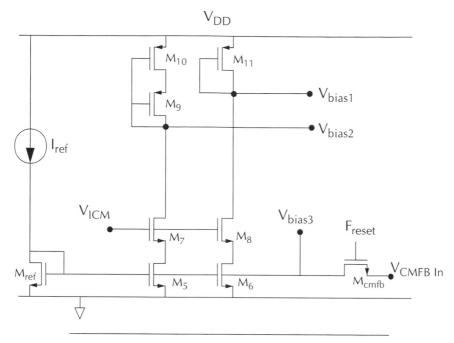

Figure 7.12: Bias generation circuitry for the OTA (shared between all OTA stages)

mirrors the current reference, and transistors M_5, M_6, M- and M_8 also mirror the reference current for use in generating bias for the PMOS current source loads. M_7 and M_8 help to match the current being mirrored into the PMOS current sources with the current flowing in the tail source of the OTA by matching the V_{ds} across M_5 and M_6 to the expected V_{ds} across the tail source (this is critical for matching the currents through the devices – the extremely poor λ of the process means I_{ds} depends strongly on V_{ds}). The tail current source, formed from transistors M_{tail1} and M_{tail2}, is split into two parts so that half of the current source can be used for common mode feedback during the evaluation phase of operation (during the reset phase, both M_{tail1} and M_{tail2} are connected to the reference voltage). The use of only half the transistor in the common mode feedback loop brings down the loop gain, and helps ensure stability. It also helps protect the circuit from charge injection and ringing from turning on and off the reset switch M_{cmfb}.

7.3.1. Optimizations

Almost any wireless system is designed with the expectation that the received signal will be small. The loss in signal power suffered from transmission through the air usually means that the signal seen by the antenna is significantly smaller than the baseband circuitry (the A/D) can detect. Invariably, amplifiers must be inserted in the receive path of the signal in order to gain the signal up to a detectable level. At the time of this writing, a typical A/D converter might expect an input signal that had been amplified to a maximum range of about +/- 1 volt or even larger. However, sustaining such a large signal swing, especially in a switched-capacitor circuit, requires extra power and area. Also, slewing may begin to hinder the speed of operation for the circuit, and headroom limitations in the OTA may begin to cause distortion or clipping unless a wide-output swing topology is chosen (e.g. a folded cascode). Therefore, it is clearly advantageous to apply only enough gain to the signal so that it meets the minimum power level so that the A/D can accurately resolve it. Adding more gain than this minimum amount can help relax the accuracy requirements for the A/D, but might incur all the potential headaches just outlined. Therefore, in a somewhat backwards design approach, the signal swing at the input of the A/D has been chosen to just meet the minimum detectable signal given the estimated offset characteristics of the A/D (allowing, of course, for some extra margin). The entire receive path has thus been designed for a maximum signal swing of +/-250 mV (differentially, -0.5 V to +0.5 V).

Figure 7.12 shows the relative sizing of the devices used to implement the OTA; the design process that yields the device sizing shown proceeds as follows:

Recall that $g_m = 2I_D/(V_{gs} - V_T)$ for an MOS transistor. Therefore, in order to achieve maximum g_m (and therefore, maximum speed) for a constrained I_D, the current density[1] of the device, $V_{gs} - V_T$, should be minimized without forcing the transistor into the subthreshold region of operation. For the OTA used in the receive chain, a $V_{gs} - V_T = 150$ mV has been chosen. Once the current density has been fixed, the current, given by:

$$I_D = \frac{\mu C_{ox}}{2}\frac{W}{L}(V_{gs} - V_T)^2$$

becomes proportional to the transistor's W/L ratio. Increasing the width of the input devices certainly improves the g_m of the OTA, but unfortunately the input and output parasitic capacitances also increase – resulting in loss of feedback factor and an increased load that needs to be driven by the OTA. The following equation shows τ as a function of the input device sizes without taking the dependence of part of C_{in} on W/L.

$$\tau = \frac{C_L}{G_m f} = \frac{C_L(C_I + C_S + C_p + C_{in})}{C_I[\mu C_{ox}(W/L)(V_{gs} - V_T)]}$$

---

1. Current density may be somewhat of a misnomer here since $V_{gs} - V_T$ does not have the units of current per unit area. It does, however, refer to the fact that the $V_{gs} - V_T$ of a MOS transistor is proportional to the square root of the drain current, I_D, divided by the W/L ratio of the device.

By taking the increased self-loading of a large device into account, the above becomes:

$$\tau = \frac{C_L(C_I + C_S + C_p + WLC_{ox})}{C_I[\mu C_{ox}(W/L)(V_{gs} - V_T)]}$$

Taking the derivative with respect to W, and setting the result equal to zero, results in a W/L that corresponds to a local minimum for τ [Conroy94]. The result of the optimization performed for this process is the two 300/0.8 micron input devices presented in Figure 7.12. The desired current level in the input devices can now be determined. From this analysis, a total tail current of 2 mA has been chosen. The sizes of all other devices in the circuit are simply chosen to meet the headroom required from the output signal swing. Finally, the tail current source has been designed with a non-minimum channel length in order to increase its output impedance (and therefore, the low frequency common-mode rejection ratio).

The designs discussed in the previous sections have been used to implement a small four-stage VGA. Each stage introduces a gain of either 0 dB or 3 dB into the signal path – resulting in a net controllable gain of 0 to 12 dB in 3 dB increments. The C_I and C_S used are 400 and 200 fF each respectively, and the transistor sizes are as shown in Figure 7.11 and Figure 7.13. The results of this prototype are documented at the end of this chapter.

SPICE simulations of the proposed VGA design (extracted from actual layout) revealed that the sample-and-hold stage met the speed and accuracy requirements dictated by the system while consuming only 10 mW of power per stage. However, when two or more stages are cascaded together, a kickback noise problem was encountered. The problem stems from the large parasitic capacitance associated with the bottom (closest to substrate) plate of the integrating capacitor C_I. At the end of each evaluation phase, C_I, and its parasitic C_{Ip}, have charge equal to C_{Vout} stored on them. As illustrated in Figure 7.7, when the next track phase arrives, C_I is switched to the input in order to sample the next value. Unfortunately, the charge stored on C_I and C_{Ip} is kicked back to the previous stage. The result is a signal dependent signal excursion at the beginning of the previous stage's settling curve. These kickback excursions severely reduce the amount of time available to the circuit to settle to its final value. A transient analysis of two cascaded stages of the VGA is shown in Figure 7.14. The clock rate has been reduced to 64 MHz because of the severe error introduced into the signal at full speed due to incomplete settling. Simulations show the VGA stages settling to the required accuracy when being clocked at 1/2 the original rate (64 MHz); however, measured test results prove this to be extremely pessimistic (90 MHz was the measured maximum frequency; see Section 7.5).

7.4. The Analog-to-Digital Converter

Of all the architectures for high speed A/D converters[1], the flash architecture is the fastest. An N-bit flash converter performs conversion simply by placing 2^N comparators in parallel and determining where, among the 2^N levels, the input lies. For a resolution as

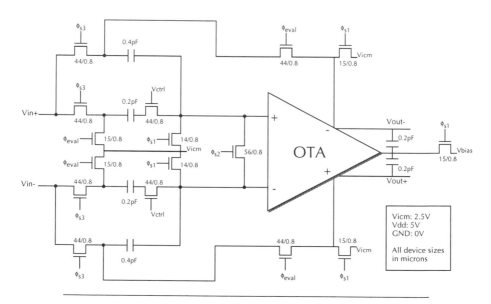

Figure 7.13: Sample-and-hold amplifier switch sizings
(see Chapter 6 for the discussion of input sampling array, and
Figure 7.11 for the OTA schematic)

small as 4 bits, a flash architecture is a feasible option for a 128 Msample/sec converter. Unfortunately, the amount of hardware and power required for an N-bit flash A/D increases exponentially as 2^N, making a full flash design less attractive because of the extremely low-power requirements of the system. However, even with a very high speed comparator design, exploration of subranging and pipeline A/D converters revealed that the 7.8 ns cycle time was insufficient for performing the extra functions required in these architectures. Furthermore, the addition of a high speed sample-and-hold amplifier (SHA) in between stages of a pipeline converter increases the power consumption due to the static current in the operational amplifier. In fact, the increase in power consumption added by the insertion of a SHA in a pipeline far outweighs the power saved by reducing

---

1. Although certain other A/D topologies, including delta-sigma and successive-approximation converters, may have high clock rates, the term "high speed" here refers specifically to a class of converters called 'Nyquist rate' converters. Unlike oversampled converters (e.g. delta-sigma), this class of converters processes one data sample per clock period, achieving the fastest possible **data** conversion rate for a given clock speed.

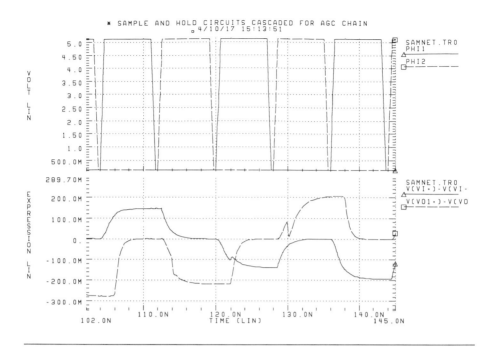

Figure 7.14: VGA stage transient analysis showing reduction in settling time due to signal-dependent kickback.

the number of comparators from 15 to 8 for a 4 bit converter[1]. The conclusion that might be drawn from this analysis is that the correct architecture to choose for the A/D converter topology is a flash. Indeed, for a stand-alone 4-bit CMOS converter operating at 128 Msamples/sec, the flash design would almost certainly be the architecture of choice. However, by merging the A/D with the VGA, further improvements are possible. The final A/D design for the CDMA receiver pipelines one of the four bits, resulting in a 1-bit to 3-bit pipeline converter. This design is depicted in Figure 7.15. Since the receiver uses a discrete-time AGC immediately before the A/D converter, each sample-and-hold amplifier in the AGC looks exactly like the interstage gain amplifier of a pipeline A/D. Therefore, the hardware to perform most of the pipelining function is already in place and could be used by the A/D at no extra cost. The AGC consists of several stages of gain, some of which must be small (for a reasonably fine gain resolution). Therefore, the interstage gain in the pipeline can be made from multiple sample-and-hold amplifiers, allowing the

---

1. Note that for higher resolution converters, the addition of a pipeline stage might very well reduce the overall power consumption of the A/D since the number of comparators increases exponentially with the number of bits, N.

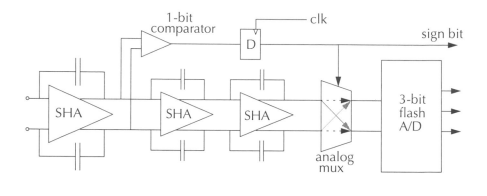

Figure 7.15: Modified 1 bit- 3 bit pipeline A/D architecture

function of generating of a residue (input signal minus analog version of MSB output) to be pipelined itself. It was found that by placing two stages of 3 dB gain (and hence two stages of delay) in between the first comparator and the remaining 3-bit flash A/D, the residue generation function could be split into two stages, thus permitting enough time to generate a residue at 128 MHz.

The receive path is fully differential; by pipelining a single bit of the A/D, the first stage (MSB) comparison becomes essentially a detection of the polarity of the incoming signal. Therefore, the 1-bit MSB subtraction function in a regular pipeline can be changed to a polarity switch (just cross-connected pass gates) based on the results of the first comparator. This concept is illustrated in Figure 7.16, and represents an absolute value function on the input signal, or analog rectification. This seemingly insignificant difference between 1/2 range subtraction and polarity swapping has a surprising effect — since the magnitude of the signal has been maintained, the interstage amplifiers of the pipeline can sustain the gain compression described previously without any detrimental effect. In other words, since the absolute value function must be performed by the AGC control block in any case, the polarity flip does not add a non-linearity into the AGC feedback loop — allowing the negative feedback to reject any gain variation in the amplifiers.

It should be mentioned here that the topology in Figure 7.15 is not a true pipeline A/D. A true 1-bit to 3-bit pipeline topology would require a gain of exactly 2 after the first comparator stage. The two SHA's are really two variable gain stages (0 or 3 dB each) in the AGC. Therefore the total interstage gain between the two comparator stages can be either 0 dB (no gain), 3 dB, or 6 dB (gain of 2) based on the RSSI (received signal strength indicator) detected by the AGC. Indeed, since a fair amount of gain droop can be tolerated in the system, the finite DC gain of the OTA's may cause the total gain to be even less than 3

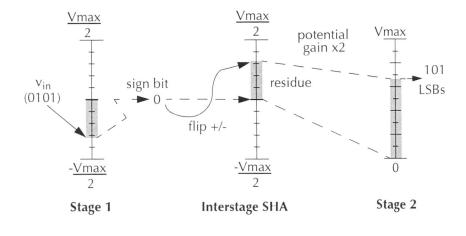

Figure 7.16: Functional illustration of the 1b-3b pipeline converter

or 6 dB. Fortunately, the variability of the interstage gain does not have a significant impact on the design. Since the feedback effect of the AGC control loop tries to force the signal at the output of the AGC to be of constant power, the reference ladder for the last stage of the A/D is fixed. The interstage sample-and-hold amplifiers between the two stages may be set to provide no gain at times. When this situation exists, offset requirements of the first stage comparator are identical to those of the last stage. However, when 6 dB of gain exists between the two stages, the signal (and the allowable offset) at the first stage is half that seen in the last stage. As with a traditional pipeline, overranging comparators and digital correction could be implemented to fix any error made by excessive offset in the first stage. However, as will be described below, the design of the single sign-bit comparator has a smaller offset characteristic than those of the other comparators. Thus, even when the signal swing at the input to the first stage of the pipeline is half of the full range, the converter performs without incorrect codes even without the benefit of digital correction in the overall pipeline.

7.4.1. Comparator Design

The heart of any analog to digital conversion circuit is the comparator. The comparator performs the quantization function of the A/D by making a decision about the input signal relative to some fixed reference. The most basic of converters, the flash A/D, simply compares the input to each of the 2^N possible discrete values between the maximum allowed input and zero. Each of the comparators decides if the input is larger or smaller than one of the reference levels, resulting in a code of '1's and '0's that can be decoded into a digital word representing the input's magnitude. In this situation, the factor that

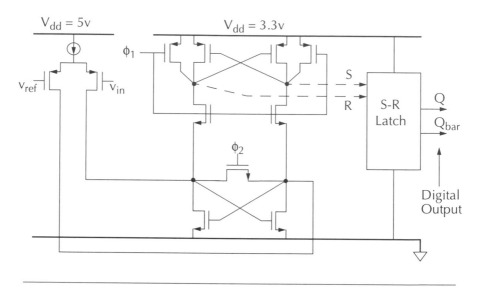

Figure 7.17: High-speed, low-offset comparator (after [Yin92])

fundamentally limits the resolution of the A/D converter is the minimum resolvable signal that the comparator can correctly make a decision upon. This key parameter is determined by several different characteristics of the comparator design, including speed of regeneration, overload recovery, and random offset. The primary difficulty in achieving high accuracy in a comparator is the inherent random offset associated with any differential structure. A differential pair ideally has a differential output of zero only when the inputs are exactly equal. However, since the devices and parasitics on each side of the layout cannot match each other exactly, this is never the case. The input offset voltage, V_{os}, is the input voltage that compensates for these non-idealities, and brings the outputs to zero. V_{os} is dependent on process variations, temperature gradients, and geometry mismatches of the design. For a simple MOS differential pair, V_{os} is given by [Gray84]:

$$V_{os} = \frac{\Delta V_T}{V_T} + \frac{(V_{gs} - V_T)}{2}\left[\frac{\Delta(W/L)}{W/L} + \frac{\Delta Load}{Load}\right]$$

An error in the comparator decision can be made if $|V_{in} - V_{ref}|$ is less than or equal to $|V_{os}|$. As always, a trade-off exists between speed and accuracy. Therefore, the high speed throughput required of the A/D converters in the receiver prohibits the use of most techniques for combating the offset problem (e.g. offset cancellation, large preamplification). However, a comparator topology has been proposed that achieves a very good combination of speed and accuracy and is shown in Figure 7.17 (after [Yin92]).

This comparator topology has a single-ended input and therefore needs to be modified for the receiver's differential signal path. This could be accomplished through the use of a capacitive input sampling network included immediately preceding the comparator, such as the one shown in Figure 7.18. Much like the sample-and-hold amplifiers, the network captures a sample of charge proportional to v_{in+} - v_{icm} on the sampling capacitors during ϕ_2. When ϕ_1 closes, one side of C_{S+} is shorted to v_{ref+}, forcing the other side of the capacitor to v_{in+} - v_{icm} - v_{ref+}. Similarly, the other side of the circuit generates v_{in-} - v_{icm} - v_{ref-}. Since v_{icm} is a common mode voltage, the differential pair generates a differential output current proportional to $(v_{in+} - v_{in-})$ - $(v_{ref+} - v_{ref-})$. Therefore, the inclusion of this additional circuit has a two-fold purpose. Not only does it provide an efficient differential-to-single ended conversion for the comparator, but it also performs a rudimentary sample-and-hold operation that might be useful if the A/D is not preceded by an active sample-and-hold circuit. The extra sampling operation provided by the input network of Figure 7.18 is certainly not needed in the receiver, since the input signal has already been sampled and held. Furthermore, a network that depends on large, very precisely matched capacitors becomes very unattractive in a standard single-poly digital CMOS process like the one available. Not only does the addition of large capacitors severely affect the input bandwidth of the A/D, it also becomes very costly in terms of area when the process used does not support a second layer of polysilicon. As discussed in Chapter 6, precision capacitors for the receiver have been created using the dielectric between the first-to-second and second-to-third metal layers. These capacitors consume so much area, and incur so much unwanted parasitics, that the prospect of creating $2 \cdot 2^N$ such capacitors for an N-bit flash converter is extremely unattractive. Therefore, a modification of the original comparator has resulted in a design capable of accepting a fully differential input, and comparing it to a differential reference without the benefit of an input sampling network.

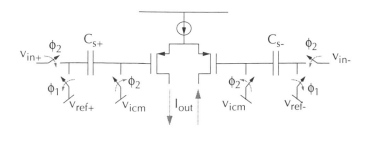

Figure 7.18: Capacitor-based input-sampling network

As shown in Figure 7.20, the original PMOS differential pair input has been replaced with the modified double-differential pair topology shown in Figure 7.19b. The output of the circuit is given by the difference between the two currents I_{out_r} and I_{out_l}:

$$I_{out} = \left(\frac{-g_m}{2}\right)((v_{in-} - v_{ref-}) + (v_{ref+} - v_{in+})) - \left(\frac{-g_m}{2}\right)((v_{ref-} - v_{in-}) + (v_{in+} - v_{ref+}))$$

Therefore, if the bias currents match exactly, the output current is proportional to $(v_{in+} - v_{in-}) - (v_{ref+} - v_{ref-})$, which is exactly the output generated with the input sampling network. Notice that this topology does not use the differential pairs in the traditional manner – v_{in+} is not being compared directly with v_{in-}. A more standard topology is shown in figure 6.19a. The differential output current is identical to that of the output of the modified circuit, but there is a significant difference between the two. Suppose a full range input signal is being compared to the maximum reference (i.e., $v_{in+} - v_{in-}$ is at its maximum value, as is $v_{ref+} - v_{ref-}$). In order for the differential pairs to stay in the linear range of operation (i_{out} proportional to differential input voltage), the $V_{gs} - V_T$ bias for the input devices must be larger than one-half the differential input. In other words, if the devices in the differential pair were biased with a $V_{gs} - V_T = 200$ mV, and 500 mV were to be applied across the input terminals, then one side of the differential pair would be completely shut off, while the other would carry the entire current from the tail source. Unfortunately, when trading off bandwidth for power, $V_{gs} - V_T$ should be kept as small as possible to achieve the highest bandwidth for the smallest power (since $g_m = 2I_D / (V_{gs} - V_T)$, this is really a matter of maximizing g_m for a given current level).

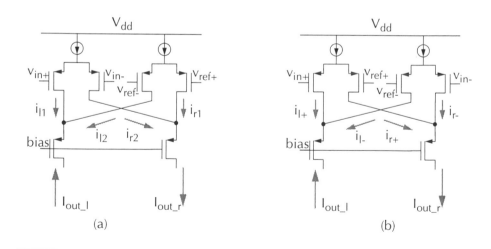

Figure 7.19: Fully-differential input sampling network. (a) Standard topology; (b) Modified topology employed by the 1b-3b pipeline A/D. Note the change in inputs

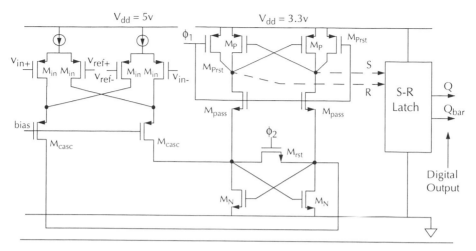

Figure 7.20: Final high-speed differential comparator

Therefore, making the V_{gs} - V_T bias as large as the full scale input voltage swing adversely affects either the speed or the power consumption of the circuit; likewise, a large V_{gs} -V_T also reduces the available headroom on the input stage. However, the proposed design in Figure 7.19b does not have this problem. Since v_{in+} is paired with v_{ref+}, and v_{in-} is paired with v_{ref-}, both differential pairs will be well within their linear range of operation even when both the input and the reference are near their maximum values. While it is true that the differential pair will become unbalanced when the input differs greatly from the reference voltage, this is of little importance because a comparator is only interested in the signal when it is near the crossing point with the reference.

The final comparator topology is shown in Figure 7.20. As described above, the input stage performs a conversion from a doubly-differential voltage to a single differential current. PMOS cascode transistors have been added to the input stage to both improve the output impedance, as well as to eliminate the Miller effect at the input. The cascode devices also help to improve the regeneration speed of the comparator since the parasitic capacitance on the drains of the cascodes can be reduced by undersizing the devices (the input devices must necessarily be larger to achieve a higher g_m). Therefore, the input stage can be regarded as a single-pole voltage-to-current converter with a high output impedance. With this in mind, the operation of the rest of the circuit can now be analyzed. The comparator requires two clock phases, as shown in Figure 7.21. The two phases must be non-overlapping, a fact which is common for many A/D converters, but is especially important for the operation of this particular topology. The clock generation circuitry is slaved off the same multiphase clock generator that is used for the sampling demodulator (Section 6.5.2).

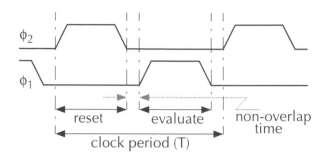

Figure 7.21: Non-overlap clock phasing for the comparator.

The first phase of operation can be called the reset phase and is illustrated in Figure 7.22. The output nodes, which go to the s-r latch, are reset to V_{dd}. During this time, the s-r latch maintains its last value. Nodes 1 and 2 are shorted together by an NMOS switch with on resistance Ron, providing a good overload recovery for circuit[1]. Although M_1 and M_2 are connected in a positive feedback loop, the loop gain is forced to be less than 1 by the reset switch – making M_1 and M_2 appear to be diode-connected. However, the input

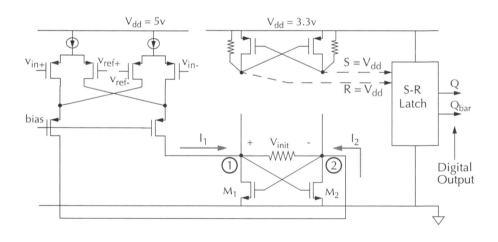

Figure 7.22: Comparator during reset phase of operation ($\phi_1 = 0$, $\phi_2 = 1$)

---

1. Overload recovery refers to a comparator's ability to recover from evaluating a sample and then correctly evaluate a subsequent value.

stage is acting as a differential current source, forcing current down M_1 and M_2. The differential current flowing down the two devices will create a differential voltage at the two drain nodes, with each device acting as a $1/g_m$ impedance load. Therefore, the voltage shown as v_{init} in the figure is proportional to the comparator's input voltage by the g_m ratio of the input devices to the NMOS latch devices (M_1 and M_2). The function of the reset phase is to provide a good recovery from the previous sample and to set up an initial voltage proportional to the input (but smaller) as the starting point for regeneration of the NMOS latch. At the end of the reset phase, clock ϕ_2 goes to zero. After the falling edge of ϕ_2, there is a brief period of time during which both ϕ_2 and ϕ_1 are low. This 'non-overlap' period is typically regarded simply as a separator between phases of operation that guarantees that one phase will not interfere with the operation of another. However, at such a high frequency of operation, the brief non-overlap period represents a significant percentage of the clock period, and really can not be sacrificed as an idle period. The state of the proposed comparator during the period when both clocks are low is shown in Figure 7.23. The reset switch across nodes 1 and 2 has been opened. Therefore, the differential current being provided by the input stage no longer sees two diode-connected loads. If the voltage v_{init} is small, then M_1 and M_2 are biased at about the same level, and act as a pair of NMOS active loads. If the initial voltage, v_{init}, is larger than a few tens of millivolts, then the positive feedback of the latch will force the two nodes to split apart further towards either supply. However, if v_{init} is small (corresponding to an input voltage very close to the reference voltage) then the currents flowing from the input stage will see the output impedance, r_o, of M_1 and $M2$. Therefore, the voltage gain of the circuit changes from $A_{v1} = -(g_{m-input}/g_{m\_1,2})$ to $A_{v2} = -(g_{m\_input}r_{o\_1,2})$. While the first term, $g_{m\_input} / g_{m\_1,2}$ is actually less than one, the second term represents an amplification of

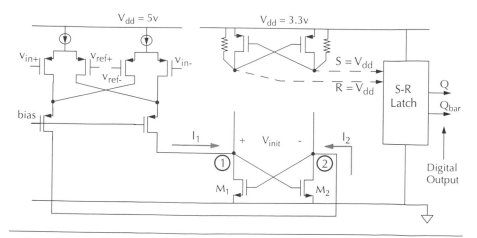

Figure 7.23: Comparator during non-overlap period ($\phi_1 = 0$, $\phi_2 = 0$)

the signal. The voltage across nodes 1 and 2 will begin to approach $v_{in} \bullet A_{v2}$ until either the non-overlap period ends (and the evaluation phase begins), or the voltage between the two nodes becomes large enough for the positive feedback of the latch to begin to re-generate it. It is not uncommon for comparator topologies to include a pre-amplifier (in the form of a differential pair) immediately before a latch because the preamp reduces the input referred offset of the latch by its gain.[1] The proposed topology folds the preamp and the latch together by using the gain achieved during the non-overlap period to re-duce the input referred offset of the latch once the evaluation period begins. The evalua-tion phase of operation begins when ϕ_1 rises (Figure 7.24). The two PMOS switches that originally shorted the output nodes to the supply are opened, and the two NMOS switches connecting the PMOS latch to the NMOS latch are closed. The resulting circuit is simply two inverters connected in positive feedback (a full latch), which regenerates the signal to near-digital levels. The true digital s-R latch has been added to both bring the final out-put to a full digital swing, as well as to hold the comparator's last output value during the reset phase. The final combination of doubly-differential input stage, high-impedance cascode, non-overlap period preamplification, and s-R latch results in a comparator that simultaneously has low offset, good overload recovery, little kickback noise, low power, and high conversion rate [Lynn95].

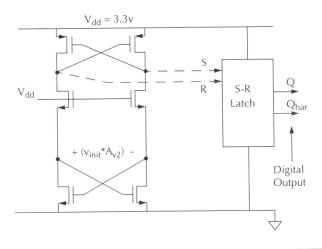

Figure 7.24: Comparator at the beginning of the evaluation phase ($\phi_1 = 1$, $\phi_2 = 0$)

---

1. Analog latches usually have a fairly poor offset characteristics because the positive feedback regeneration grows exponentially with time. Therefore, a small signal will start out with a slow rate of change, resulting in poor offset performance.

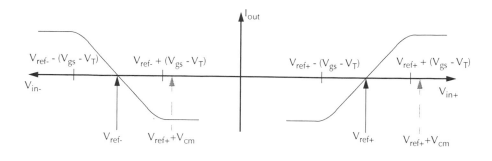

Figure 7.25: Effect of common-mode shift on the modified input stage

7.4.2. A/ D Optimizations

The previous section explained the advantages achieved from using the modified input
stage shown in Figure 7.19b over the standard double-differential stage of Figure 7.19a.
With its limited linear range, the differential pair limits the ability of the circuit in
Figure 7.19a to compare a large input to a large reference voltage; the improved design in
Figure 7.19b avoids this because it places the comparator switching point exactly in the
middle of the differential pair's linear range. However, if the common mode voltage of
the input is different from the common mode voltage of the reference, then the switching
point no longer falls exactly in the middle of the differential pair transfer function, and
results in an offset in the comparison point. This concept is illustrated in Figure 7.25.
The differential output current is given by the sum of the two currents given by the
curves shown in the figure. However, if the difference in common mode voltage shifts the
comparison point outside the linear region of the circuit (as shown by the light grey arrows in Figure 7.25), then the differential output near the switching point will be zero,
regardless of the input's value. In other words, the modified topology of Figure 7.19b
works well even for large signal swings, but begins to fail if the common mode difference
between input and reference varies significantly from zero. Therefore, the V_{gs} - V_T bias of
the input devices must be chosen so that the circuit will remain linear during the largest
expected common mode excursion. Unfortunately, the input offset voltage of a MOS differential pair is given by:

$$V_{os} = \Delta V_t + \left(\frac{V_{gs} - V_t}{2}\right)\left[\frac{\Delta(W/L)}{W/L} + \frac{\Delta Load}{Load}\right]$$

Clearly, the offset due to geometry mismatch in the devices increases with V_{gs} - V_t. Furthermore, the g_m of the input stage is $2I_D$ / (V_{gs} - V_T). Not only can the g_m for a given
current can be maximized by reducing V_{gs} - V_T, but the offset can be reduced in this

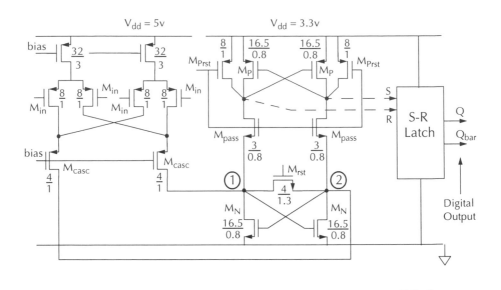

Figure 7.26: Comparator device sizing (microns)

manner as well. Therefore, the bias on the input devices must be carefully designed in order to have a minimal $V_{gs} - V_T$, while still ensuring some robustness against common mode excursions. To this end, an input $V_{gs} - V_T = 200\text{mV}$ has been chosen. In order to meet the stringent power budget allocated for the A/D, a total tail current of 40 mA was allowed (10 mA flowing though each input transistor). Using the drain current equation for an MOS transistor,

$$I_D = \frac{\mu C_{ox}}{2}\frac{W}{L}(V_{gs} - V_T)^2$$

the W/L sizing for the input devices can be determined. The choice of $8\mu/1\mu$ geometry input PMOS's leads to an estimated worst case input offset of 25-30 mV for the comparator, just below half of an LSB:

$$\frac{\Delta(W/L)}{W/L} \cong 0.05, \; \Delta V_t \cong 10\text{mV}$$

Once the input device dimensions are determined, the cascode transistors cannot be far behind, else the nondominant pole at the source of the cascode device becomes significant. Therefore, the two cascode transistors have been designed to be half the size of the input devices in order to reduce the capacitance contributed to the NMOS latch. The sizes of the various transistors used in the comparator are shown in Figure 7.26. The NMOS latch transistors should be made as large as possible in order to achieve a faster regeneration speed by increasing current drive. However, the parasitic capacitance at the two nodes 1 and 2 is partially determined by the width of transistors M_N. Therefore, an optimum transistor width can be found from the time constant equation:

$$\tau = \frac{C_{1\,total}}{g_{mN}} = \frac{C_{parasitic} + \alpha W_N}{\sqrt{2I_N\mu C_{ox}(W_N/L)}}$$

where α is the capacitance per unit width contributed to node 1 (or node 2) by the two M^N transistors. Optimization of the above yields:

$$W_N = \frac{C_{parasitic}}{\alpha} = 16.5\mu m$$

The initial voltage from which regeneration will start is determined by:

$$V_{init} = -\left(\frac{g_{min}}{g_{mN} - (2/R_{onrst})}\right)(V_{in} - V_{ref})$$

Therefore, the R_{on} of the NMOS reset transistor must be made large enough to ensure a sizeable initial signal voltage, but small enough to provide a good overload recovery during the reset phase. Since the primary purpose of the reset switch is to drop the positive loop gain of the latch below one, the design equation becomes:

$$\left(-g_{mN}\frac{R_{onRst}}{2}\right)\left(-g_{mN}\frac{R_{onRst}}{2}\right) < 1$$

The above evaluates the loop gain around the latch. By taking the square root of both sides, and substituting for g_m and R_{on}, the relative sizing of transistors M_N and M_{rst} can be determined:

$$\frac{(\sqrt{2\mu C_{ox}I_N(W/L)_N})}{2}\left(\frac{1}{\mu C_{ox}(V_{gs} - V_t)_{rst}(W/L)_{rst}}\right) < 1$$

$$(W/L)_{rst} > \frac{\sqrt{\mu C_{ox}I_N}}{\mu C_{ox}(V_{gs} - V_t)_{rst}}\sqrt{\frac{(W/L)_N}{2}}$$

For the 0.8 micron CMOS process available, the above becomes:

$$(W/L)_{rst} > \left(\frac{1}{6}\right)\sqrt{(W/L)_N} \cong \frac{2}{3}$$

A W/L_{rst} = 4/1.3 microns was chosen for the reset switch, but probably could have been more optimally designed as a smaller (i.e., lower W/L ratio) device.

Device sizes for the other transistors in the circuit have less impact on the comparator's performance. NMOS transistors M_{pass} were sized to minimize the parasitic capacitive load placed on the NMOS latch. The PMOS latch devices, M_P, were sized to match the current drive of the NMOS latch transistors as is done in any digital design (once the evaluation phase arrives, the n-latch combines with the p-latch to effectively make two cross-coupled digital inverters). The final design is shown in Figure 7.26.

It should be mentioned here that the sign bit comparator performs a comparison against a zero-valued reference voltage. In other words, this one comparator does not need four inputs, since it only determines if v_{in+} is greater or less than v_{in-}. The device geometries for this comparator all remain the same as for the comparator just described in this section, but the lack of the two extra inputs gives this comparator a better offset characteristic. It is the increased resolution of this design that allows it to be used in the 1-bit to 3-bit pipeline. Since the two interstage SHA's between the sign bit comparator and the 3-bit flash A/D represent a potential gain of 6dB, the sign bit comparator must have more accuracy in order to resolve a smaller signal.

SPICE simulations of the proposed comparator design (extracted from actual layout) showed the comparator performing at the speed and accuracy requirements for which it was designed. Figure 7.27 shows a transient output from a comparator simulation. The top panel of Figure 7.27 shows the input samples to the comparator. The horizontal line across the panel corresponds to the reference voltage. The second panel shows the digital output of the S-R latch switching every time the input crosses the reference voltage. The bottom panel shows the two non-overlapping 128 MHz clocks used to control the comparator. Simulation of the entire A/D converter, although too cumbersome to include here, was consistent with the simulation of each individual comparator.

7.5. Results

Three chips have been fabricated in a standard HP non-double-poly 0.8 micron digital CMOS process. The first die consists of the proposed 3-bit flash A/D structure. A diagram of the 3-bit flash converter layout is shown in Figure 7.28, and a close-up diagram of one of the comparators is shown in Figure 7.29. It is important that the comparator layout be as symmetric as possible in order to prevent an increase in V_{os} due to geometry mismatch. Substrate and well contacts can be seen surrounding the circuits in the bottom half of the figure, forming rectangular guard rings to collect as much substrate noise current as possible. The signal flows from the bottom of Figure 7.29 to the top where the S-R latch can be seen (sandwiched between two bypass capacitors). It is also worth noting that the clock distribution is horizontal across the circuit (and not directly above any transistors) so that inductive and capacitive coupling between the clock and the signal path (which flows vertically through Figure 7.29) is minimized.

A diagram of the 3-bit A/D test chip (which includes a sample-and-hold circuit) is shown in Figure 7.30. Large on-chip bypass capacitors can be seen distributed throughout the die. The signal enters the chip from the left-hand side of Figure 7.30 and the digital out-

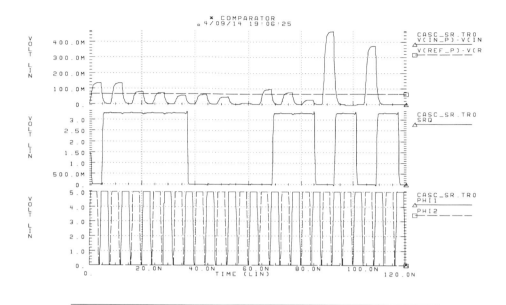

Figure 7.27: Single comparator transient analysis showing correct
digital output based on comparison of input to reference.

puts exit the chip on the right of the figure. All pads on the right-hand side of Figure 7.30
are reserved for digital signals (including digital supply and clocks) and the analog and
digital supply and ground connections are completely separated on the chip (except for
the fact that the grounds must be connected through the substrate of the die).

Speed tests on this die have shown that the A/D functions up to a 150 MHz clock rate
(the goal of the design was 128 MHz). Beyond 150 MHz, the converter does not have
sufficient time to evaluate a signal and convert it to a full digital level. Total power consumption from the analog circuitry is 2.2 mW. Digital power consumption for the converter could not be determined because the vast majority of digital power consumption
on the test chip is created by the buffers driving the outputs off chip at the full 128 MHz
rate (which would not exist in a fully integrated implementation).

The second die was mentioned in Chapter 6 with regard to RF layout of a sampling demodulator; it consists of a standalone I/Q demodulator, with analog outputs to measure
distortion performance and RF analog track bandwidth. Settling performance of the OTA
could not be verified with this die; the output capacitive loading on the OTA is that of the
pad, board-level parasitics, and the scope probe: far greater than the on-chip load it was
designed to drive. Because of this, the die could not be clocked any faster than 20 MHz;
however, this was sufficient to acquire the measurement results for which it was designed:
namely, the RF track bandwidth and distortion of the sampling demodulator.

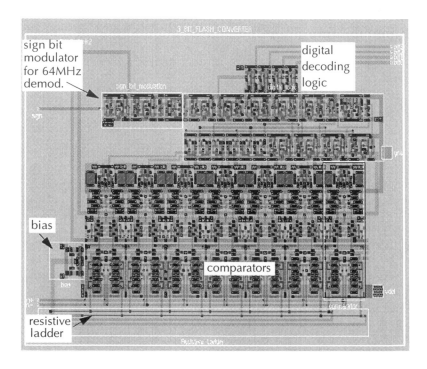

Figure 7.28: 3-bit flash A/D converter layout

The die photo is shown in Figure 7.31. Measured -3 dB point on the input track bandwidth is 1.78 GHz (compared to a simulated 2.1 GHz), and its IP_3 distortion intercept was +22 dBm (slightly higher than the simulated +20.4 dBm). To measure the distortion intercept, two equal-power 0 dBm sinusoidal input tones were injected, one at 1.020 GHz and the other at 1.021 MHz. Subsampling at 20 MHz, the third order intermodulation component at 2 MHz was measured using an HP 70000-series spectrum analyzer. These results agree well with the simulated results discussed in Chapter 6, and are well within the accuracy of the device simulation models provided by the foundry; a summary table is provided as part of Figure 7.33.

Lastly, the final die represents the entire analog front-end testchip: it includes the full 1-bit- 3-bit A/D converter, the complete 4-stage VGA chain, the sampling demodulator, and the RF LNA. Due to time constraints, the AGC control loop has not yet been designed/implemented; however, the gain is externally controllable, and can be set on the testboard. The die photo is shown in Figure 7.32. Excluding the LNA, the top half and the bottom half of the core are symmetric copies of one another; the LNA (being single-ended for power considerations) is placed as far away from the high-speed switching path as possible. Again, the die is symmetric with respect to the in-phase and quadrature signals;

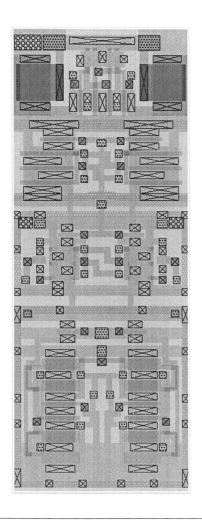

Figure 7.29: Comparator layout

the centerline of symmetry divides the I channel and Q channels of the receiver. The signal flows from left to right across Figure 7.32, and the pad ring to the right of the chip has been broken to separate analog and digital supplies.

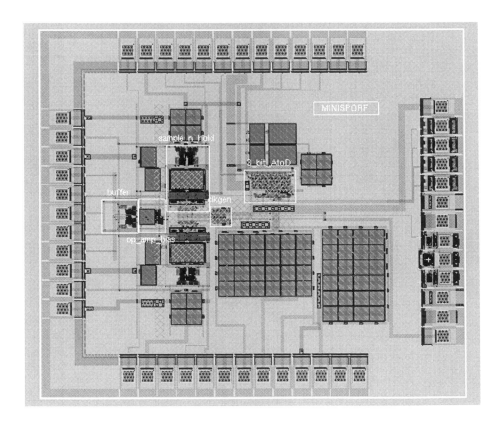

Figure 7.30: 3-bit flash A/D converter testchip

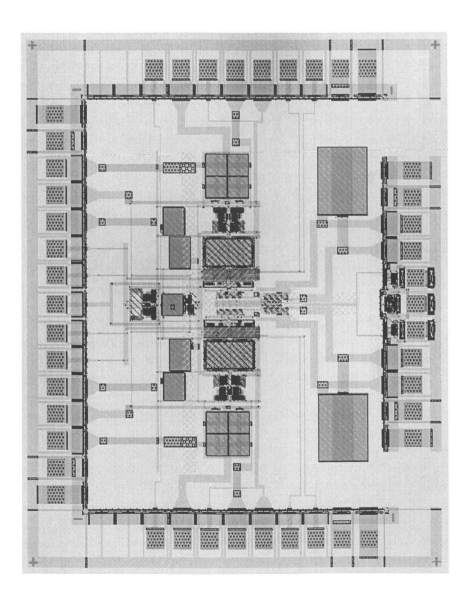

Figure 7.31: Sampling demodulator testchip

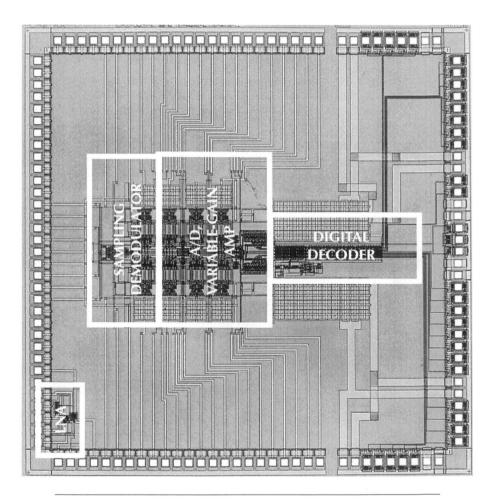

Figure 7.32: Full analog front-end receiver testchip

Parameter	Value	Parameter	Value
LNA		Baseband VGA	
Gain	11.5 dB @ 1.05 GHz	Maximum Sampling Rate	90 MHz
Noise figure	7.2 dB	Digitally Controllable Gain (4 stages)	12 dB, 3 dB/stage
Power Consumption	36 mW @ 3.3 V	Per Stage Sample & Hold Power Consumption	8.3 mW
IP3	-10 dBm		
Sampling Demodulator		A/D Converter	
I/Q Phase Balance	4°	Peak SNDR	25.6 dB
Track Bandwidth	1.75 GHz	4 bit A/D INL	0.42 LSB
Noise figure (computed)	18.1 dB	4 bit A/D DNL	0.36 LSB
IP3	22 dBm	Maximum A/D Conversion Rate	150 MHz
		A/D Power Consumption (for two converters on chip)	4.4 mW
		Total Measured Analog Power Consumption	106.8 mW

Figure 7.33: Measured results for the analog front-end receiver testchip. LNA results are from Chapter 6; sampling demod track bandwidth and IP3 are from the standalone sampling demodulator testchip results. Maximum A/D conversion rate and A/D power are from the standalone A/D testchip results.

The RF chip-on-board high-speed test board described in the appendix to Chapter 6 was used to test the integrated analog receiver chip. The final measured results for the complete receiver chain are listed in Figure 7.33, including a recap of the LNA results from Chapter 6. Beyond the LNA, the key test results from this die were the operating frequency and linearity of the entire baseband chain. The maximum clock frequency of the overall chain proved to be significantly lower than the 150 MHz top speed of the flash A/D converter, owing to the kickback problem described previously in Section 7.3. The maximum sampling rate at which an input sinusoid can be recovered without error at the output is 90 MHz; beyond this, loss of settling time due to the signal dependent kickback noise causes the sample-and-hold amplifiers to fail. Lastly, Figure 7.34 shows the FFT of the baseband sampling-demodulated output of an RF signal; in particular, the input was modulated 1.024 GHz carrier modulated in single-sideband by a 100 kHz sinu-

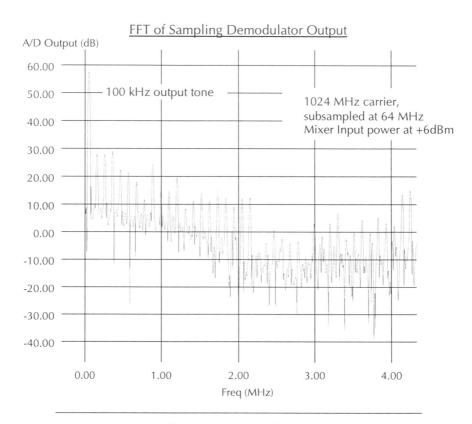

Figure 7.34: FFT of baseband sampling demodulator output. Input signal is a +6 dBm, 1.024 GHz carrier, SSB modulated by a 100 kHz sinusoid. Output is sampled at 64 MHz; distortion harmonics are consistent with that of a 4-bit A/D converter.

soidal tone, subsampled at 64 MHz. By integrating the distortion harmonics the signal-to-noise plus distortion ratio (SNDR) of the entire receiver chain has been calculated to be 25.6 dB (or slightly over four bits).

Lastly, as a full demonstration of the baseband transmit modulator and analog front-end, the modulator's output was used to drive the semicustom, broadband single-sideband RF modulator, utilizing the down-up method described in Chapter 5. No off-the-shelf band-pass filter possessed sufficient bandwidth to pass the full transmit spectrum; a cascade of lowpass-highpass filters was used to implement crude bandpass filters. Due to this, the carrier frequencies were limited to 700 MHz; above this, the noise admitted into the system due to the imprecise nature of the filters made testing above this frequency unreliable. To accommodate the digital acquisition system at the output of the A/D converters,

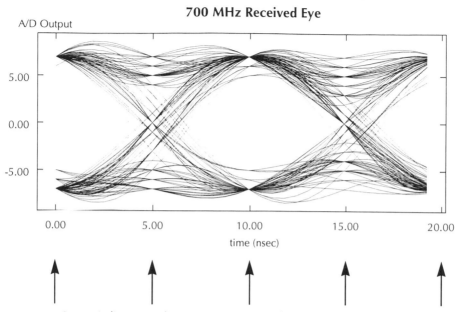

Arrows indicate sampling instants (20 MHz). The arcs in-between were generated from the A/D outputs using an ideal DAC interpolator on a workstation.

Figure 7.35: 700 MHz received eye diagram.

the system spread-spectrum chipping rate was set to 20 MHz. Figure 7.35 shows the received eye diagram for a transmitted PN sequence, which is the pilot tone by itself. The eye is clearly open, with quite a bit of both noise immunity and jitter margin, and the 15 levels of the A/D converter are quite distinct at the sampling instants.[1] Furthermore, the off-time sampling point is clearly saturating the A/D converter, as compared to the true 30% excess bandwidth eye diagram of Figure 3.2. To test data integrity, a sliding correlator test was applied to the received PN sequence; a single strong peak should be readily apparent, if no data corruption was incurred. Indeed, a single peak appears at a value equal to 7•32768 (maximum A/D output times the full PN sequence length). Thus, at least for this test, the raw bit-error rate is significantly lower than 1 part in $3•10^5$ bits.

---

1. The sampling rate here was 20 MHz; the arcs in between the sample instants was reconstructed after digital acquisition using an ideal DAC interpolator on a Sun workstation.

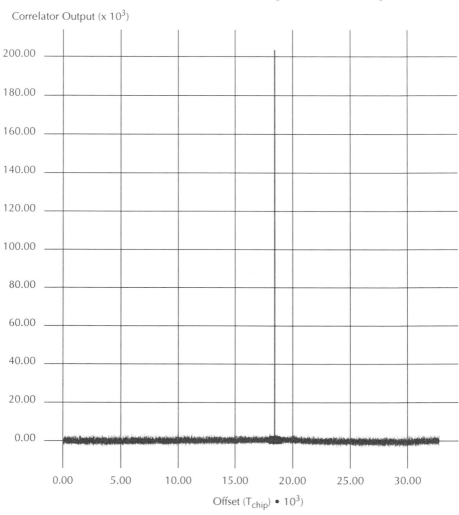

Figure 7.36: 700 MHz received data, sliding correlation test

REFERENCES

[Conroy94]   C. Conroy. *A High-Speed Parallel Pipeline A/D Converter Technique in CMOS.* Ph.D. Thesis, U.C. Berkeley, February 1994.

[Gray84]   P.R. Gray and R.G. Meyer. *Analysis and Design of Analog Integrated Circuits, 2nd Edition.* New York: John Wiley and Sons, 1984.

[Gray89]   P.R. Gray. Course Notes for EECS 290Y, Analog-Digital Converter Technologies. Department of EECS, U.C. Berkeley, Berkeley, CA. Spring 1989.

[Lin90]   Y-M. Lin. *Performance Limitations of High-Resolution Video-Rate Analog-Digital Interfaces.* Ph.D. Dissertation, Memorandum No. UCB/ERL M90/55, U.C. Berkeley, Berkeley, CA, June 19,1990.

[Lynn95]   L. Lynn. *Low-Power Analog Circuits for an AII-CMOS integrated CDMA receiver.* M.S. Thesis, U.C. Berkeley, Berkeley, CA, 1995.

[Uehara93]   G. Uehara. *Circuit Techniques and Considerations for Implementation of High-Speed CMOS Analog-to-Digital Interfaces for DSP Based PRML Magnetic Disk Read Channels.* Ph.D. Thesis, U.C. Berkeley, Berkeley, CA, 1993.

[Yin92]   G. Yin, F. Op't Eynde, and W. Sansen. "A High-Speed CMOS Comparator with 8-b resolution." *IEEE Journal of Solid-State Circuits*, Vol. 27, No. 2, pp. 208-211. February 1992.

8 The Receiver: Baseband Spread-Spectrum Digital Signal Processor

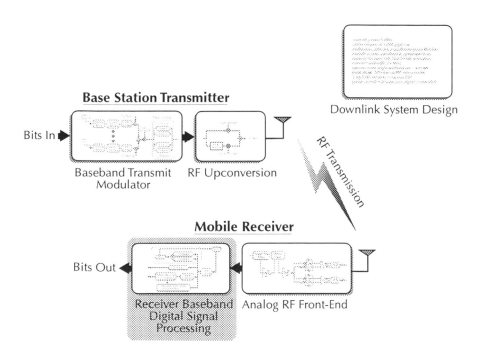

Base Station Transmitter

Bits In ►

Baseband Transmit Modulator

RF Upconversion

RF Transmission

Downlink System Design

Mobile Receiver

Bits Out ◄

Receiver Baseband Digital Signal Processing

Analog RF Front-End

Last, but certainly not least, the digital signal processing required to recover the original transmitted data needs to be addressed. From the output of the receiver's analog front-end, two 4-bit, 128 MHz interleaved streams are emitted: the only processing that has been done so far is to bring the signal to baseband, and A/D converting it into a digital stream. Even beyond sheer data recovery, the issues of timing recovery and carrier frequency recovery have not been addressed, nor has the implementation of such receiver functions as channel estimation, multipath combining, and adjacent cell detection. None of these issues are touched in the analog domain; instead, they are relegated to the baseband signal processor that will be discussed in this chapter. A more detailed description of the prototype circuit implementation can be found in [Stone95]; the discussion in this chapter and the next is intended to focus on the specific system and circuit design tradeoffs that were made, with the next chapter focusing on the design of the core functional block: the digital matched-filter correlator.

As mentioned before, this partitioning of the receiver architecture – an "A/D front-end" coupled with a highly complex digital signal processor (DSP) – has resulted in a low-power solution for the analog aspects of the receiver. However, this may place the power burden in the digital domain: the 128 MHz programmable DSP's of today consume literally watts of power, and it might be surmised that a custom application-specific integrated circuit solution would be comparable. Fortunately, this is not true; from the classic equation for digital switching power, $P_{digital} = CV^2f$, reduction of the supply voltage V is the key to low power consumption, owing to the quadratic dependency. With an application-specific integrated circuit, architectural techniques can be applied to preserve the operating frequency f, potentially with some expense in the switching capacitance C, and yield tremendous net gains in power efficiency [Chand92]. In particular, the total power consumption of the baseband receiver DSP is a mere 46.5 mW: far lower than if a programmable DSP were employed to provide the same functionality. In general, digital techniques are better suited than analog in creating circuits that can operate at high speed on a low supply voltage – a critical aspect of low-power design.

8.1. Receiver DSP Architecture

In Figure 8.1,[1] the block diagram of the receiver DSP is shown. As expected, it implements all of the functionality delineated in Chapter 3 for the spread-spectrum receiver: baseband timing recovery, adjacent-cell detection, channel estimation, and data recovery. Interestingly, this architecture is particularly amenable to low-power implementation: all of the functional blocks run in parallel, with little interaction between them. In particular, this parallel nature allows the input data (clocked from the analog receiver front-end at 128 MHz = $T_{chip}/2$) to be immediately decimated to 64 MHz (at T_{chip}). The 128 MHz data stream can be considered to be two interleaved substreams at 64 MHz, one consisting of "on-time" data, and the other of "off-time" data. The "on-time" samples are fed to the data recovery, channel estimation, and adjacent-cell scan blocks; the "off-time" samples are fed to the delay-locked loop for timing recovery, as will be discussed below.

However, one crucial block is missing: frequency recovery. Due to the need for minimal phase noise in the sampling demodulator architecture, frequency "pulling" of the baseband clock oscillator driving the demodulator is highly undesirable; it would render the entire demodulating front-end unusable, from excessive jitter. Digital methods of fre-

---

1. The implementation prototype for the DSP implements a subset of this block diagram; see section 8.6.

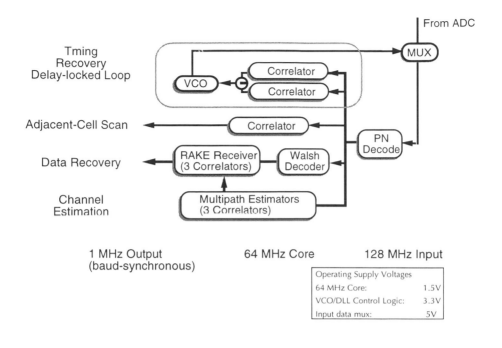

Figure 8.1: Receiver DSP block diagram

quency compensation are thus called for. The most straightforward – an all-digital phase-locked loop – is also the costliest is power and complexity, given the 128 MHz input rate and the required ancillary hardware: read-only memory lookup tables, interpolators, and digital multipliers. Instead, the high frequency accuracy and stability of crystals are leveraged: the worst-case frequency offset for the crystal oscillators employed by this system is 25 kHz. Compared to the 64 MHz spread-spectrum chipping time, the phase is varying extremely slowly, and can be compensated incoherently. As each of the functional blocks is discussed, the impact of incoherent recovery will be made apparent.

Lastly, it is also apparent from Figure 8.1 that the key block required in the receiver DSP is the matched-filter correlator. If a low-power implementation can be developed – operating at 64 MHz on as low of a supply voltage as possible – a low-power spread-spectrum receiver is thus produced. As the viability of the entire "A-to-D" front-end architecture hinges upon power efficiency in the DSP back-end, the entirety of the next chapter is devoted to examining the implementation of the matched-filter correlator.

8.2. Timing Recovery and Adjacent-Cell Detection

Arguably, the most difficult and important problem that faces the receiver DSP is that of baseband timing recovery: without accurate and consistent timing lock, all of the complex channel estimation and multipath combining strategies are meaningless. As discussed in Chapter 3, the pseudonoise (PN) pilot tone, with its strong autocorrelation properties, is used to synchronize the receiver with the base station. The implementation of the acquisition-and-tracking loop used to track the pilot tone in the receiver is the subject of this section.

8.2.1. Coarse Acquisition

First, when the mobile unit is turned on, detection of the base station and coarse timing acquisition must be performed in the receiver DSP. By "coarse", it is meant that the receiver's internal PN sequence is synchronized to within ± 1 T_{chip} of the base station's transmitted PN sequence. Effectively, coarse acquisition consists of stepping through each possible phase of the PN sequence, correlating at each phase, until a significant correlation output is detected. The full 32768-length correlation is not necessary at each PN phase; from simulations, 1024-length correlations are sufficient to minimize the probability of false lock, as well as minimize the time to acquisition (from Chapter 3, the worst-case synchronization-lock time is 128 milliseconds). To achieve this lock time, four parallel correlators are used, searching four separate PN code phases simultaneously. A diagram of the coarse acquisition block is shown in Figure 8.2; algorithmically, it can be described as:

1. Initialize PN generator and reset all correlators. Each correlator is time-offset by 1 T_{chip} with respect to one another, i.e., $correlator_0$ will correlate over $phase_n$, $correlator_1$ over $phase_{n-1}$, $correlator_2$ over $phase_{n-2}$, and $correlator_3$ over $phase_{n-3}$.
2. Correlate at this phase for 1024 cycles.
3. Latch output of each correlator.
4. Check each the magnitude of each output to see if it exceeds the lock threshold. If any do so, a positive lock has been achieved, and go to stage 6. Otherwise, continue to stage 5.
5. Reset the correlators to search for the next group of 4 PN phases. Stall the PN generator 4 four clock cycles. This will shift $correlator_0$ from $phase_n$ to $phase_{n-4}$, $correlator_1$ from $phase_{n-1}$ to $phase_{n-5}$, etc. Note that $phase_{-1}$ is $phase_{32767}$ (phases are modulo 32768). Go to stage 2 and repeat.
6. A positive lock has been found. Stall the PN generator a number of cycles equal to the index of the correlator that found lock; this forces the PN phase on $correlator_0$ to be the positive-lock phase for simplicity in control. Thus, if $correlator_0$ found lock, do not stall; if $correlator_1$ found lock, stall 1 cycle, etc.
7. PN phasing is set within ± 1 T_{chip} of the correct value. Switch to fine tracking mode, delay-locked loop.

Since silicon die area is at a premium, it would be a waste to expend the four correlators used in coarse-lock only for this purpose. In particular, these four correlators are well suited to perform the functions of channel estimation and adjacent-cell detection, which will be discussed later. However, some hardware multiplexing is necessary to support switching functionality; a more detailed diagram of the long correlators can be seen in Figure 8.3. The signal CMPTH-L is fed from each of the long correlators to the lock circuitry, whereas the other signals, IPQORI and Q, are used after lock for the remaining functions, such as multipath channel estimation.

8.2.2. Fine Timing Acquisition and the Digital Delay-Locked Loop

After coarse lock has been achieved, the sampling point must then be moved as close to the maximum eye opening as possible, and this point must be tracked with respect to receiver movement, channel fading, and baseband oscillator frequency offset between transmitter and receiver. Shown in Figure 8.4a is the 1024-length correlator output as a function of time offset, for an ideal received transmit pulse (with zero offset being perfect synchronization). As expected, it is simply proportional to the 30% excess bandwidth raised-cosine transmit pulse. By locking two separate correlators E_{early} and E_{late} on either side of the peak (at $\pm T_{chip}/2$ relative to the center), an estimate of the peak location can be determined by the difference between the two; the symmetric nature of the pulse shape is critical. If the sampling point is early, then the difference in magnitude output

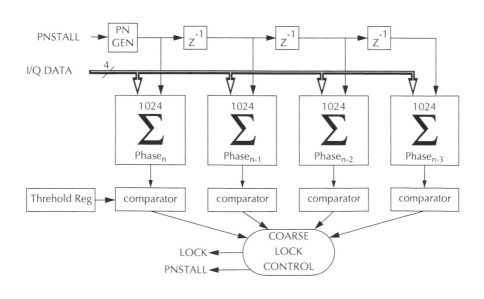

Figure 8.2: Block diagram from coarse lock acquisition. The hollow arrow indicates 4-bit I/Q pair.

between E_{early} and E_{late} will be negative, driving the sampling point forward. If the magnitude difference is positive, then the sampling point will be driven in the opposite direction (Figure 8.4b), thus following any change in the received signal timing.

This forms the basis of the fine-tracking delay-locked loop (DLL); however, its implementation, especially with regard to the analog front-end, needs to be carefully considered. In the analog front-end, the 128 MHz baseband oscillator is used to drive the demodulator, subsampling the 1.088 GHz received signal. However, the same oscillator must also be used to drive the baseband DSP: "shifting" the sampling point in the analog domain for timing recovery becomes extremely difficult, since it will have dramatic effect on the RF demodulation characteristics. Fortunately, another solution is available. As discussed in Chapter 6, recovery of both in-phase and quadrature signals requires two separate 128 MHz sampling switches, interleaved in time to effectively form a 256 MHz converter. As far as the analog front-end is concerned, designation of one stream as "in-phase" and the other as "quadrature" is purely notational; at its output, the two streams are effectively equivalent. The receiver DSP, on the other hand, can take advantage of this symme-

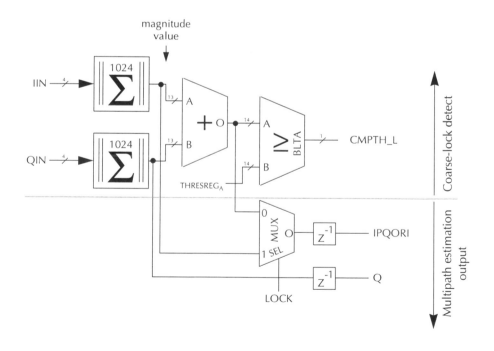

Figure 8.3: Block diagram of a multipath correlator.

try: with respect to the 256 MHz data stream, the only needed fact is that any two consecutive samples will be 90° out-of-phase with respect to one another, regardless of which demodulator actually generated which sample (Figure 8.5).

Thus, for the purposes of timing recovery, the data stream is effectively oversampled at 4X with respect to the chipping rate; no real need exists to shift the sampling phase in the analog domain. Instead of driving an analog voltage-controlled oscillator, the delay-

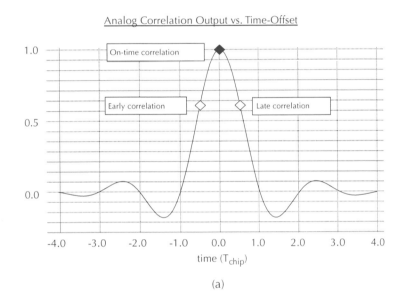

(a)

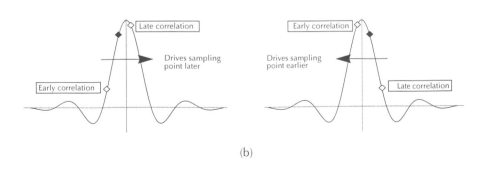

(b)

Figure 8.4: (a) Correlator outputs as a function of time offset (t=0 is synchronized). (b) Early/Late phase sensing in the delay-locked loop.

locked loop instead drives a 256 MHz multiplexer at the input of the receiver DSP; it is this multiplexer that decides which sample is "in-phase" or "quadrature", and which sample is "on-time" or "off-time", as shown in Figure 8.5. The resulting granularity in sampling is 2 nsec ($T_{chip}/8$) in the worst-case, which is more than sufficient from system-level simulations.[1] One comment needs to be made on notation: for the discussion below, the multiplexer outputs are designated I_{on}, Q_{on}, I_{off}, and Q_{off}, being the I/Q on-time and I/Q off-time samples respectively. It is the off-time samples that are used to drive the DLL, since these represent the early/late samples of Figure 8.4.

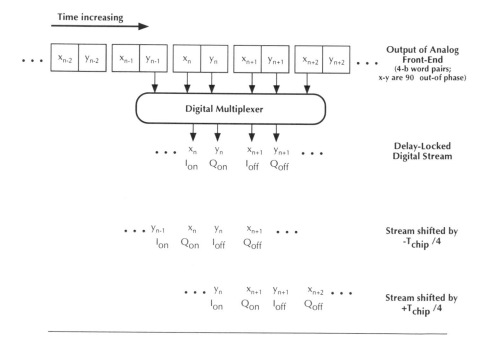

Figure 8.5: Mapping of analog front-end outputs into the four delay-locked digital data streams. The multiplexer "shifts" the samples appropriately.

1. It is this $T_{chip}/8$ sampling error that gives rise to the timing jitter specification mentioned in Chapter 3. Also, since the sampling-point adjustment is thus fixed, the DLL tracking is akin to step adaptation in adaptive equalizers, with the error signal being the difference between the E_{early} and E_{late} correlators.

The two basic building blocks of the DLL need to be addressed: the phase detector, which consists of the E_{early} and E_{late} correlators, and the input multiplexer, which serves the role of a variable delay. First, a functional block diagram of the phase detector can be seen in Figure 7.8. One of the 1024-length correlators receives an I_{off}/Q_{off} pair, while the other receives an I_{off}/Q_{off} pair which has been delayed by T_{chip} (the late-correlator input is simply the early-correlator input delayed by 1 T_{chip}). Both correlators extract the required pilot tone magnitude from the off-time data, over a period of 1024 samples (16 μsec), and then latched. From the recovered early and late pilot tone energies, a comparison is made between the absolute value of the difference between E_{early} and E_{late} ($|E_{early}-E_{late}|$) to a threshold register. If the comparator output CMPTH3 is greater than the value programmed in the threshold register, then the timing error is sufficient to warrant a shift in the sampling phase; the difference ($E_{early}-E_{late}$) is then used to determine the direction of the shift (Figure 8.4). By setting an "acceptable" threshold on the difference through CMPTH3, needless limit-cycle oscillation in the sampling phase is hence prevented when E_{early} and E_{late} have similar values.

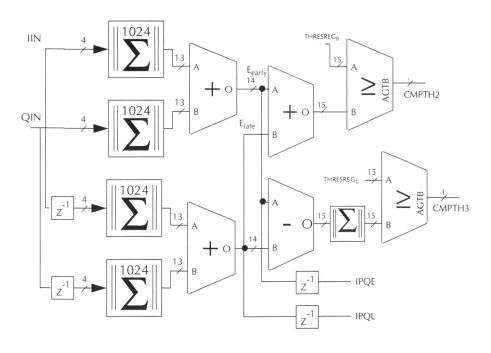

Figure 8.6: Block diagram of DLL phase detector. CMPTH3 indicates whether the system clock's phase should be adjusted. CMPTH2 indicates whether the received tone energy is large enough to stay in lock.

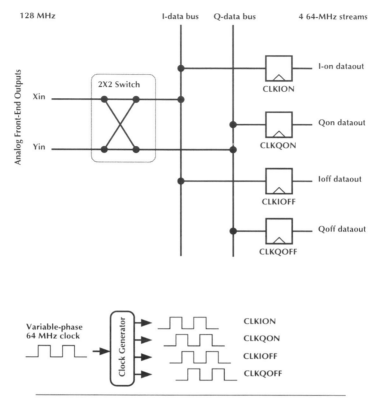

Figure 8.7: DLL VCO (data multiplexer) block diagram. All datapaths above are 4-bits wide.

Second, the variable delay element for the DLL – the front-end I/Q data multiplexing – is shown in Figure 8.7. By selecting which sample is in-phase or quadrature, as well as establishing which sample is designated as being "on-time", it can shift the effective sampling point by the desired resolution of $T_{chip}/4$. From an architectural standpoint, there are four registers at the input to the DSP, designated I_{on}, Q_{on}, I_{off}, and Q_{off}. The two 4-bit samples from the analog front-end are first passed through a 2x2 digital switching network (implemented with true digital multiplexers), designating which sample is "I" or "Q". The "on-time" or "off-time" designation is then made by controlling which register is clocked, latching the appropriate sample into the appropriate register. By multiplexing the samples with this gated clocking technique, shifting the on-time sampling point by $\pm T_{chip}/4$ becomes a matter of extending or shortening the clock period for one cycle, instead of inserting or deleting samples from the input stream. This is desirable from both a complexity and power standpoint, since the control only needs to act on the clock (a 1-bit stream), instead of attempting to shift 4-bit sample words from register to register.

As always, an architectural description is far simpler than its implementation; the $T_{chip}/4$ adjustment of 4 nsec implies that the clock generators and their control logic must run at 256 MHz, resulting in a key problem with critical path delays. Likewise, when the clock phase is shortened by $T_{chip}/4$ to perform the timing adjustment, care must be taken that this shorter cycle time (12 nsec; Figure 8.8) does not impact the correlators, which are designed to operate with a 16 nsec cycle time. Thus, reducing the clock phase by $T_{chip}/4$ in actuality is implemented as an extension in time by $3T_{chip}/4$, with the obvious loss of a single sample. Since the time-phase adjustment can happen only once every 1024 samples, a potential loss of 1 sample every 1024 is acceptable, resulting in SNR degradations of 0.01 dB in the timing recovery block and 0.13 dB in the data recovery block. However-er, the PN and Walsh generators cannot lose a clock edge, otherwise the entire DLL scheme would be useless; separate clocks are necessary for the correlators and the PN/Walsh gen-erators, which are designed to meet the shorter 12 nsec cycle time.[1]

The implementation of the variable-phase clock generator is shown in Figure 8.9, along with its timing diagram. It consists of two stages, with the first stage generating a vari-able-phase 64 MHz clock (designated CLKX), and the second stage generating the four re-quired clock phases (CLKION, CLKIOFF, CLKQON, CLKQOFF) relative to CLKX. It is

Variable-phase 64 MHz Clock

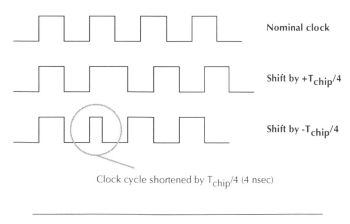

Nominal clock

Shift by $+T_{chip}/4$

Shift by $-T_{chip}/4$

Clock cycle shortened by $T_{chip}/4$ (4 nsec)

Figure 8.8: Variable-phase 64 MHz clock
(adjustable by $\pm T chip/4$)

---

1. This requirement is not severe; the PN/Walsh generators are fairly straightforward, with only a few gate delays in the critical path; their implementation is the same as that described in Chap-ter 4 for the transmit modulator integrated circuit.

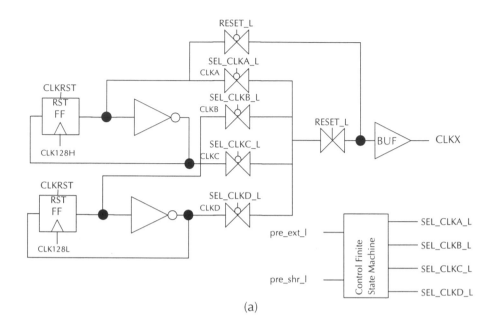

(a)

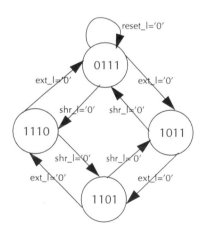

State transition diagram for the four SEL_CLK signals. Each state bit corresponds to a control bit for the clock synthesizer's pass-gates. Thus:
bit3(msb)=SEL_CLKA_L
bit2=SEL_CLKB_L
bit1=SEL_CLKC_L
bit0(lsb)=SEL_CLKD_L.

pre_ext_l, if asserted, extends the phase by $T_{chip}/4$.
pre_shr_l, if asserted, shortens the phase by $T_{chip}/4$

ext_l = latch (pre_ext_l)
shr_l = latch (pre_shr_l)
reset_l = 0 forces return to 0111 state.

(b)

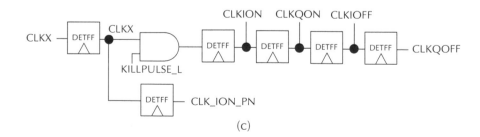

(c)

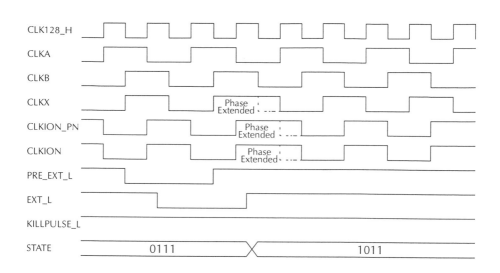

Timing diagram for reducing the clock's phase.

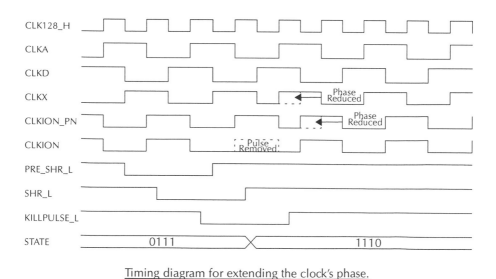

Timing diagram for extending the clock's phase.

(d)

Figure 8.9: (a) 64 MHz variable-phase clock synthesizer.
(b) State machine to control clock phase.
(c) 4-phase clock generator (derives the four required clocks from CLKX)
(d) Timing diagrams

important to note that when the clock phase is being reduced by $T_{chip}/4$, CLKX actually does shorten by that amount; the second stage's KILLPULSE gate control line guarantees that this shortened pulse is not fed to the datapaths on the chip or to the sample latching. As described above, the clock for the PN and Walsh generators cannot lose this pulse, and thus their clock is derived from CLKX directly, without any gating. The four DETFF blocks in the second stage are true dual edge-triggered flip-flops [Afgh91] being clocked by the $T_{chip}/2 = 128$ MHz input clock; by latching the gated version of CLKX on both edges of the 128 MHz clock, a $T_{chip}/4$ delay line is formed, generating the four clock phases.

Lastly, the implementation of the input data multiplexer is shown in Figure 8.10; it differs significantly from Figure 8.7 in two ways. First, the "on-time" and "off-time" designation registers have been moved before the I/Q designation mux; this is to avoid critical path problems with data changing at the mux input and a clock edge attempting to latch the output data 4 nsec later. Second, to simplify clocking in the remainder of the chip, all of the samples are then time-aligned to the I_{on}-clock phase, allowing the remainder of the chip to run at $1/T_{chip}$ (64 MHz) synchronous with CLK-ION. This resynchronization is performed by a sequence of three latches; the first latch R_1 synchronizes I_{on} with Q_{on} and I_{off} with Q_{off}, the second latch R_2 synchronizes all four together, and the third latch R_3 resynchronizes the data with the correlator's clock (CLKIONI-5X). One might question the need for the third latch. The reason for this is because no clock skew can be allowed between CLKION, CLKQON, CLKIOFF, or CLKQOFF; thus, to equalize the loading on all clocks, the CLK-ION signal must be buffered before it is sent to the remainder of the chip. However, the buffered output CLKIONI-5X is skewed with respect to the original CLK-ION. Hence, the final latch is used to resynchronize the four output clocks with the master clock for the entire chip, which is CLKIONI-5X.

8.2.3. Carrier-Offset Compensation

As discussed in Chapter 6, for the crystals employed in this system, there is a worst-case frequency offset of 25 kHz between the transmit and receive RF local oscillators. This frequency shift results in a continuous, undesirable phase change of the received baseband signal, i.e., the baseband received constellation will be slowly "spinning," in addition to the phase change seen by channel variations and receiver motion. With a 25 kHz carrier offset, a 64-long correlation spanning 1 μsec will see a shift of approximately 9° during the correlation length, which is acceptable for DQPSK data recovery purposes: the spin is strongly correlated from sample-to-sample; so long as its impact on the actual correlation output is small, it will tend to differentially cancel. However, for timing recovery, all of the blocks described above rely on a 1024-long correlation: 144° worst-case across its 16 μsec span, and intolerable for purposes of accurately estimating the channel peak. The impact of such spin can be seen in Figure 8.11[1]; instead of a ramp function, the peak becomes degraded, since the phase change results in a cancellation of the desired component of the correlation. Likewise, the phase is constantly rotating in the correlator output; instead of a flat phase output, the constellation is effectively "spinning". Shorter correlations are not possible, due to subcorrelation noise effects and reduced processing gain – extremely detrimental to the critical function of timing recovery. A method of compensating this phase change must be developed.

For the magnitude estimates needed in both timing recovery and adjacent cell detection, the compensation algorithm is based on the fact that (after PN multiplication), the pilot-tone sequence component of the received data is constant, with additive high-frequency noise from other users, multipath, and neighboring cells. Effectively, the actual hardware correlator functions as a lowpass filter. Thus, provided that the high-frequency noise component can be damped away, the following Cauchy-Schwartz (in)equality holds:

$$\left| \sum_{i=1}^{1024} a_i \right| \approx \sum_{j=1}^{m} \left| \sum_{i=1}^{n} a_i \right|$$

where n•m is equal to 1024, the correlation length. In essence, since the desired component is constant, the magnitude of the sum is equal to the magnitude of the partial sums without phase rotation. Under phase rotation, the partial (shorter) sums will have a far smaller error, and thus yield an accurate, long magnitude correlation output without being impacted by the carrier offset. Empirically, it was found that there was no difference between n=64, m=16 and n=128, m=8; either of these values was optimal. The case of n=64, m=16 was chosen for convenience, since n=64 corresponds to the system processing gain. A plot of the modified correlator output is shown in Figure 8.12, and a block diagram of the magnitude correlator structure is given in Figure 8.13.

It should be mentioned that the magnitude estimators are implemented with an L_1 norm (given as ($|I| + |Q|$)) in the complex-valued I/Q space, instead of a conventional Euclidean L_2 norm $\sqrt{I^2 + Q^2}$. In Figure 8.14, the impact of this choice is shown – the unit circle in the complex plane has been deformed into a cycloid, with a worst-case magnitude error of +3dB. The reason for this was complexity minimization; a true Euclidean L_2 norm would have been extremely difficult to implement in a low-power fashion without consuming excessive die area; from simulation, the performance degradation due to this was found to be negligible.

8.2.4. Adjacent-Cell Detection

Lastly, the problem of adjacent-cell detection for mobile-assisted soft-handoff must be addressed. It is quite similar to the problem of coarse timing acquisition: the pilot-tone energy from adjacent cells must be recovered. By maintaining a single sliding correlator,

---

1. Instead of simply showing the correlator output at the dump instant, the entire integration process is shown. Ideally, it should be a ramp; instead, there is significant degradation due to the phase rotation caused by the frequency offsets. It is important to note that for a length-64 correlation, there is no significant deviation; for a length-1024 correlation, the error becomes severe.

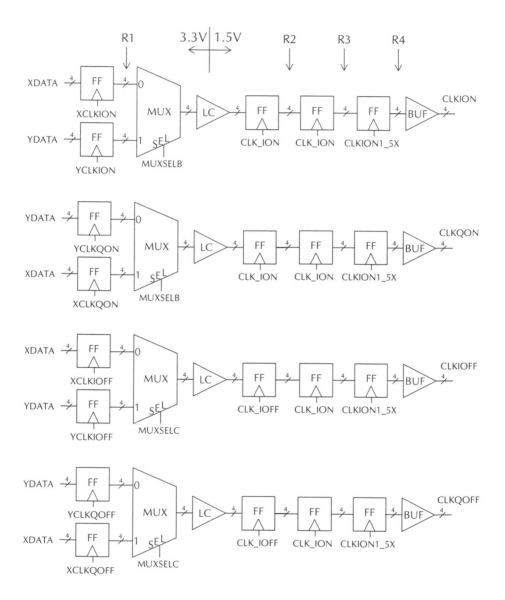

Figure 8.10: Block diagram of the input multiplexer.
(LC= level converting buffer; XDATA, YDATA are the 4-bit words
from the analog front-end).

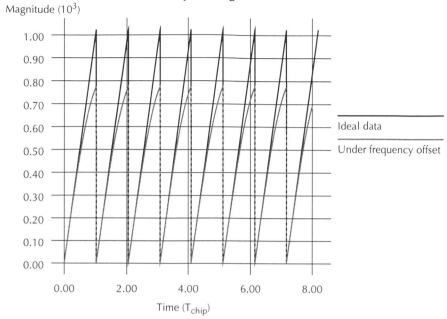

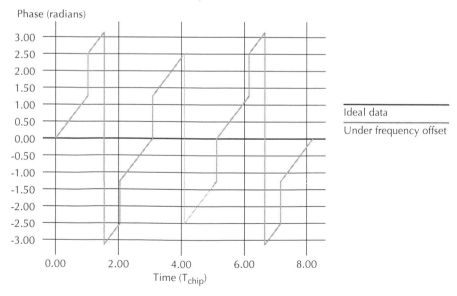

Figure 8.11: 1024-long correlation output, ideal and under 25 kHz carrier frequency offset.

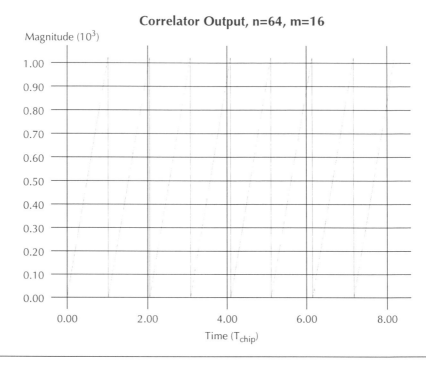

Figure 8.12: Compensated magnitude correlator output, n=64, m=16. Peak value is 1022, compared to a nominal output of 1024.

correlating over a 1024-long interval and changing its phase by 1 T_{chip} every correlation interval, a search can be made for PN-sequence correlation peaks due to nearby cells (Figure 8.15). Peaks within $\pm 4\,T_{chip}$ of the current cell's PN correlation timing peak must be discounted, since they are most likely caused by multipath effects. As described in Chapter 3, with the long PN sequence, there exists a sufficient number of code-phase offsets to accommodate a K=7 cellular reuse pattern, even accounting for PN code synchronization errors between cells. Since only the magnitude of the correlation is necessary to determine if an adjacent cell is present, the carrier-offset compensation technique described in the previous section can be employed.

However, once this sliding correlator has found an adjacent cell (decided when its 1024-long correlation output has exceeded a certain threshold CMPTH4), the issue remains of how to store where these peaks occurred, so a soft-handoff can be performed quickly by the mobile. In essence, some kind of "register file" would be needed to store the PN code phase; three related methods have been devised to perform this.

- Seed storage. After detection of an adjacent cell, the PN code phase of the sliding correlator is no longer adjusted. When the code phase of the primary PN generator (locked to the current cell's pilot tone) reaches the all-one's state in its shift register, the state of the sliding correlator's shift register is parallel-stored into a register file, along with the measured strength of the adjacent-cell pilot and the Walsh code counter value at that instant[1]. The advantage of this is clear; very little digital hardware is necessary to store the phase relationship of the adjacent-cell pilot. Shifting the primary PN generator over to the new phase is simple: when it reaches the all-one's state in its shift register, the adjacent-cell's seed stored in the register file is parallel-loaded into its shift register, thus forcing it into coarse synchronization with the adjacent cell. The disadvantages are that fine-lock acquisition is still necessary, and moving "back" to the code phase of the original cell is difficult, as its code phasing is lost.

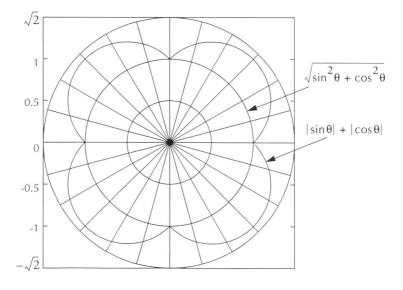

Figure 8.13: Graph comparing the absolute value L_1 norm against the true L_2 norm.

---

1. The storage of the Walsh code is required to guarantee that both PN and Walsh are synchronized when handoff occurs. Otherwise, a large time penalty may be incurred to recover the Walsh start point (all one's state in the shift-register PN generator).

- Multiple PN/Walsh generators. Instead of simply storing the seed, a second strategy would be the implementation of multiple PN/Walsh generators, all running independently. After detection by the sliding correlator, the PN and Walsh code phases are retained by storing them into another PN/Walsh generator. These secondary PN/Walsh generators can be switched-in to replace the primary generators when a handoff is necessary. This solves the problem of moving "back" to the original cell's code phase after handoff, since its code phasing has been retained. However, the issue of fine-lock acquisition latency still remains, and there is a larger penalty in hardware overhead.

- Replication of the entire receiver delay-locked loop. Lastly, the solution that overcomes all of these issues is to replicate the entire delay-locked loop to track the adjacent cells [QualC92]. Fine acquisition is accomplished by default; the tracking loop is ready to recover data the instant a handoff is called. Obviously, the hardware penalty is enormous - all of the correlators, clock generation, and control circuitry must be replicated[1].

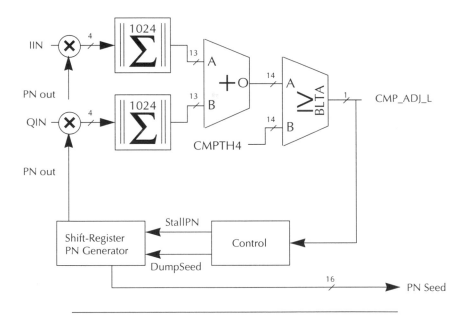

Figure 8.14: Block diagram of the adjacent-cell detector
(sliding correlator)

---

1. 0.9 cm x 0.9 cm, fabricated in a 0.8 micron HP CMOS process. The DLL and data multiplexer blocks comprise slightly over 30% of this.

Clearly, there is a tradeoff between the speed at which the soft-handoff can be performed, and the hardware required to implement such a "phase register". However, in light of the size of the prototype receiver DSP (which does not implement adjacent cell-scan), a compromise solution between simple seed storage and having multiple PN/Walsh generators would seem to be the clear solution: seed storage of the detected adjacent cells, with two parallel PN/Walsh generators to allow quickly jumping back to the original cell's PN phase. Although replication of the entire receiver DLL provides the best performance, it is not practical unless a more advanced fabrication technology were to become available.

8.3. Channel Estimation

Like the problem of timing recovery, channel estimation relies on the receiver being able to accurately recover long correlation streams of the pilot tone. As discussed in Chapter 3, by utilizing correlators with their pilot PN sequences synchronized to on-time, $+1\ T_{chip}$, and $+2\ T_{chip}$ time-sampling instants, the multipath channel profile may be estimated. Since partial correlations are used, a moving-average lowpass filter is can be used to filter out the subcorrelation noise (Figure 3.9). However, the issue of frequency compensation remains: as before, even with subcorrelations as short as 1024 in length, the worst-case phase change is still intolerable. Unlike the timing recovery case, phase information is critical to the operation of a multipath combining receiver structure. A simple magnitude estimate will not suffice; a method to eliminate the frequency offset of the oscillators must be developed.

The channel estimation algorithm is shown schematically in Figure 8.16; it is akin to the magnitude estimator of Figure 8.15, in that correlations are taken over length 64, 1 microsecond chunks with a small phase error due to the frequency offset, and the result of the correlations then averaged over 16 microseconds to statistically minimize the noise in the shorter correlations. However, instead of averaging the magnitude of the length-64 chunks, the complex division between the first and second correlations, and the first and third are averaged. By taking the complex division, this provides not only the relative amplitude of the multipath arrival to the primary arrival, but also provides the phase difference between the two. Thus, frequency-offset compensation can be achieved, while preserving the slowly-varying multipath phasing information that is desired.

Figure 8.17 shows a simulation of this estimation technique, and a comparison to the case where a simple 1024-long correlator is used. A three-path channel is assumed, with tap weights $(1.0, 0.8e^{j\pi/5}, 0.2e^{-j4\pi/5})$. Likewise, a 25 kHz oscillator offset between the transmitter and the receiver is intentionally introduced as a sinusoidal modulation at the input to the receiver block, and a three-tap MA filter is applied after the 1024-long output to minimize the noise effects due to subcorrelations (Chapter 3). The magnitude and phase estimates are shown for this estimation technique, as well as the case where the simple correlator is assumed. The results for this technique are quite good; even for an oscillator offset of 25 kHz, the worst-case estimation error (on the weakest multipath arrival) is only 2 dB in amplitude, and 12° in phase.

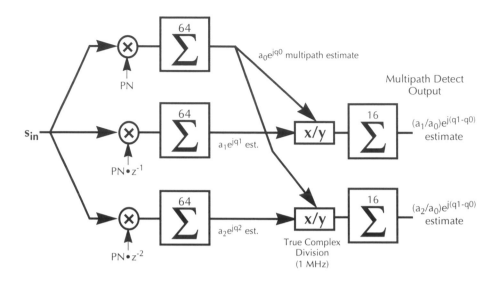

Figure 8.15: Block diagram of multipath estimator.
At the output of the length-64 correlators, carrier offset is still present. The
complex division removes the effect of the carrier, while preserving the relative
phasing and amplitude between the first and second and first and third
arrivals. s_{in} is fully complex-valued, being equal to (I_{in}, Q_{in}). Likewise, all
thick arcs in the block diagram are complex-valued.

8.4. Data Recovery

Lastly, data recovery must be discussed. Given that the sampling instant is being tracked
by the timing recovery delay-locked loop, and that reliable multipath profiles can be gar-
nered by the channel estimation loop, what remains is the multipath combining structure, along with the DQPSK data slicer block. Given that the interference-limited
combiner [Teus97] and the multi-user estimator [Verdu93, Verdu97] are still both under
development, a straight RAKE combiner will be described below, to illustrate the issues
with implementing such structures.

As described in Chapter 2, a RAKE coherently combines data recovered: so long as the relative phasing between the first, second, and third arrivals is known, it is sufficient to perform coherent combining. In some sense, the scheme can be considered to be a
differential version of the conventional RAKE algorithm [Dixon84]. The proposed modification is shown schematically in Figure 8.18; it is akin to Figure 2.14, with a division by
a_0e^{jq0} on all three RAKE fingers followed by a DQPSK slicer. The multipliers on each finger

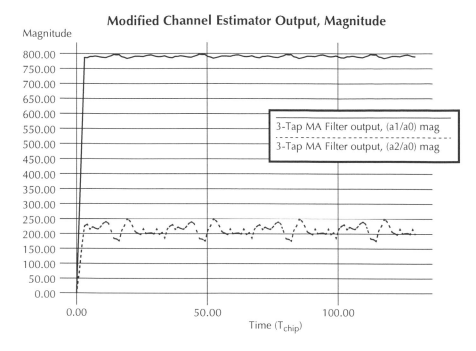

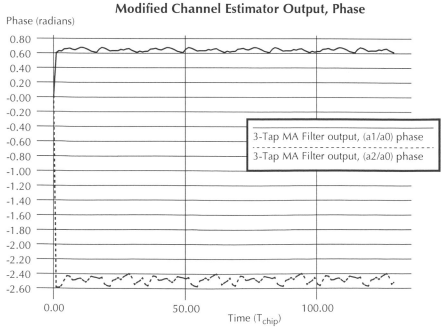

Figure 8.16: Modified channel estimator output (25 kHz frequency offset between transmit and receive oscillators).

then become the complex conjugate of the differential channel estimates described in the previous section. In the original RAKE scheme, the multiplication by $a_0 e^{-jq0}$, $a_1 e^{-jq1}$, and $a_2 e^{-jq2}$ on each finger eliminates any channel phase rotation associated with each multipath arrival: the phase is brought back to zero by the complex conjugate, and the output could be coherently decoded with a simple QPSK slicer instead. Since absolute phase information is not available, dividing all three paths by $a_0 e^{jq0}$ removes the need for absolute phase. Instead of referencing each multipath finger back to zero phase offset, each multipath finger is referenced to the phase primary arrival e^{jq0}. Thus, a true DQPSK decoder is necessary, since the combined data is phase incoherent, and dependent on the previous symbol for data recovery.

To complete the discussion on data recovery, the DQPSK decoder itself needs to be examined. In essence, the phase difference between two consecutive received symbols must be determined, and the data decoding scheme is fairly straightforward, with the caveat that a sampling demodulator is being used in the analog front-end. If the delay-locked loop slips the sampling instant, it in effect exchanges the "in-phase" designation with the "quadrature" designation; i.e., a change of $T_{chip}/4$ at baseband results in a 90° phase shift

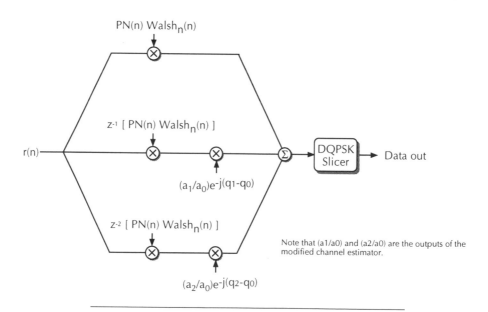

Figure 8.17: Phase-differential RAKE receiver

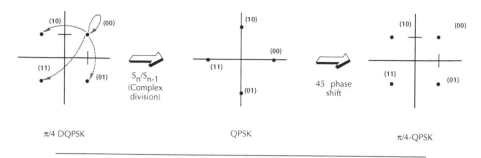

Figure 8.18: Conceptual overview of the DQPSK slicer.

at RF. Thus, the DQPSK decoder needs to implement a 90° correction in the phase if a change in sampling instant was made by the delay-locked loop. If S_n is the current output of the RAKE, and S_{n-1} is the previous output, then the slicing algorithm is as follows:

$$ \text{databits} = \text{sgn}\left(\text{Re}(\frac{S_n}{S_{n-1}}e^{j\omega\phi})\right), \; \text{sgn}\left(\text{Im}(\frac{S_n}{S_{n-1}}e^{j\omega\phi})\right) $$

where sgn() is the sign function, and ϕ is a phase correction. The division of S_n by S_{n-1} results in a complex number with a phase equal to the phase difference between the two symbols. If there is no shift in the DLL sampling instant, a ϕ of 45° translates the DQPSK constellation into an absolute phase QPSK, as shown in Figure 8.19. The slicing operation is then particularly simple, with the data bits corresponding to the sign bits of the real and imaginary parts. Although it would seem that the phase correction would require a full complex-valued multiplier, the amplitude-independence of the sgn function implies that the phase correction can be implemented with two adders, since the multiplication will be against (+1.0,+1.0), (-1.0,1.0), or (1.0,-1.0), for ϕ = 45°, 135°, and -45° respectively. However, the key question that remains is the implementation of the complex division of S_n by S_{n-1}; well-known iterative methods such as CORDIC [Thon92] are well-suited to this application.

8.5. Results

In Figure 8.20, the die photo of the receiver DSP testchip is shown. It implements a subset of the functionality described in section 7.1; the block diagram for this die is shown in Figure 8.21. In particular, the complex-valued computation (required for channel estimation and data recovery) is not present; these functions were not necessary to test the full functionality of the receiver: the data rate at the output (nominally 1 MHz) is low enough that it can be easily acquired in real-time to a Tektronix digital signal analyzer, and then analyzed on a Sun Sparcstation. Also, since no hard decision on the data is done on-chip, this also allows this chip to be used as a platform for testing soft-decoding error-correction algorithms, such as Ungerboeck coding [Unger87], as well as more combining strategies such as the interference-limited combiner [Teus97].

The die has been tested, and is fully functional up to an input clock rate of 80 MHz, consuming 46.5 mW; this value is shy of the 128 MHz clock rate it was originally designed for, due to an error with an undersized clock buffer in the clock generation circuitry of Figure 8.9. Since the vast majority off the circuitry on this die is comprised of matched-filter correlators, a standalone correlator was fabricated and tested for perfor-

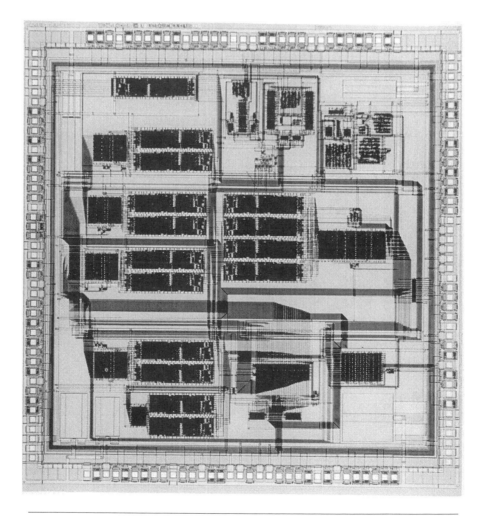

Figure 8.19: Die photo of the receiver DSP prototype testchip.

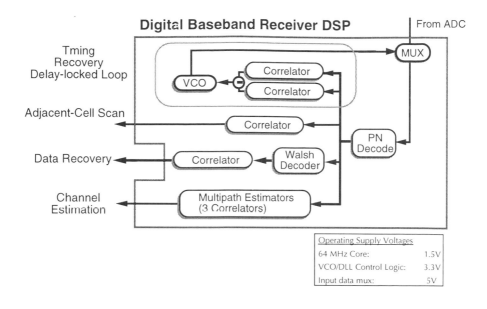

Figure 8.20: Block diagram of the functionality implemented on
the receiver baseband DSP prototype.

mance verification. As predicted by simulation, the standalone correlator was fully functional at the core operating frequency of 64 MHz at a supply voltage of 1.5 V, while consuming a mere 600 microwatts. The maximum operating frequency on this correlator was 77 MHz before it failed to function. Both the standalone correlator and the full receiver DSP testchip bear testament that high-speed digital circuitry, coupled with careful architectural design, can operate with extremely low power consumption.

REFERENCES

[Afgh91]   M. Afghahi and J. Yuan. "Double Edge-Triggered D-Flip-Flops for High-Speed CMOS
              Circuits." *IEEE Journal of Solid-State Circuits*, Vol. 26, No. 8. August 1991.

[Chand92]   A. Chandrakasan, S. Sheng, R.W. Brodersen. "Low-Power CMOS Digital Design."
              IEEE Journal of Solid-State Circuits, Vol. 27, No. 4, pp. 473-484.
              April 1992.

[Dixon84]   R.C. Dixon. *Spread Spectrum Systems, 2nd ed.*, New York: J. Wiley and Sons, 1984.

[QualC92]   *An Overview of the Application of Code Division Multiple Access (CDMA) to Digital
              Cellular Systems and Personal Cellular Networks.* Qualcomm Inc., May 21, 1992.

[Stone95]   K. Stone. *Low Power Spread Spectrum Demodulator for Wideband Wireless
              Communications.* M.S. Thesis, U.C. Berkeley, Berkeley, CA, 1995.

[Teus97]   C. Teuscher. Ph.D. thesis on Multi-Access Limited, Multi-User Combining Techniques,
              Forthcoming. U.C. Berkeley, Berkeley, CA, 1997.

[Thon92]   L. Thon. *Application-Specific Processors for Numerical Algorithms.* Ph.D. Thesis,
              U.C. Berkeley, Berkeley, CA, 1992.

[Unger87]   G. Ungerboeck. "Trellis-Coded Modulation with Redundant Signal Sets, Parts I and
              II." *IEEE Communications Magazine*, Vol. 25, No. 2, February 1987.

[Verdu93]   S. Verdu. "Multiuser Detection." *Advances in Statistical Signal Processing Vol. 2: Signal
              Detection*, pp. 369-409. Greenwich, CT: JAI Press, 1993.

[Verdu97]   S. Verdu. "Demodulation in the Presence of Multiuser Interference: Progress and
              Misconceptions." *Intelligent Methods in Signal Processing and Communications*,
              D. Docampo, A. Figueiras-Vidal, F. Perez-Gonzalez, eds, pp. 15-44.
              Boston: Birkhauser, 1997.

9 The Matched-Filter Correlator
by Ian O'Donnell

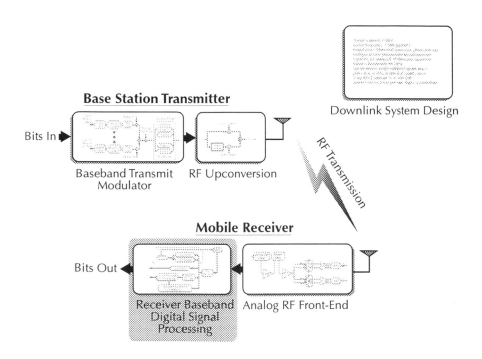

Base Station Transmitter

Bits In ▶

Baseband Transmit Modulator

RF Upconversion

Downlink System Design

RF Transmission

Mobile Receiver

Bits Out ◀

Receiver Baseband Digital Signal Processing

Analog RF Front-End

As described in the previous chapter, the viability of a 64 Mbps all-digital spread-spectrum receiver hinges upon the development of a low-power matched filter correlator. In a more general light, next-generation digital communications systems will increasingly rely on digital signal processing, given the rapid advances in silicon CMOS scaling and computational capability possible in a single custom integrated circuit. Likewise, it is also possible to simultaneously achieve extremely low power consumption, extremely attractive for portable applications. The previous chapter has already alluded to this: the fabricated correlator, in a standard digital 0.8 micron CMOS process, can achieve 64 MHz performance while consuming only 600 microwatts of power. It is the goal of this chapter to describe the design methodology which has made this possible.

One of the key developments in low-power digital design of the past several years has been the concept of architectural voltage scaling. From the equation for dynamic switching power in digital circuits:

$$P_{dyn} = C_{eff} V_{dd}^2 f$$

where C_{eff} is the effective capacitance that is being switched, f is the switching frequency, and V_{dd} is the supply voltage, it is clear that power is a strong function of supply voltage. Reducing V_{dd} becomes the key goal in low-power design; however, reducing V_{dd} also results in reducing the maximum possible operating frequency. By performing architectural optimizations such as pipelining the circuitry and the use of parallel functional blocks, throughput can be maintained while preserving low power consumption [Chandra92]. To achieve this end the supply voltage was chosen *a priori* to be 1.5V[1] – the entire correlator circuit was designed to work on this supply voltage, at a clock rate of 64 MHz.

9.1. The Correlator Design

Given the numerous times that the correlator is replicated within the spread-spectrum digital backend, it is expected to take up the largest percentage of the die area and total circuit power consumption. Thus, at a 64 MHz clocking rate, the need for both a low-power, low-area implementation is crucial. To more optimally satisfy these performance constraints a full-custom design approach was employed.

The correlator essentially functions as an accumulate-and-dump of N weighted inputs

$$Y = \sum_{i=1}^{N} W[i] \cdot X[i]$$

where X[i] is an input sample and W[i] is the weight. For the receiver DSP, the input sample from the analog front-end is 4 bits wide clocked at 64 MHz. A sign-magnitude representation was chosen; given that the received data samples are spectrally white due to the spread-spectrum nature of the system, from [Chandra94], it has been shown that a sign-magnitude correlator results in 30% lower power consumption than an equivalent two's-complement one. This is due to the nature of the weighting function W[i], which is a 1 bit stream of +/-1's corresponding to a Walsh code overlaid on a PN sequence. Since multiplication by +/-1 requires a simply toggling of the sign bit, it is extremely efficient

---

1. 1.5V was chosen due to the fact that this is a readily available battery voltage.

from a power standpoint. Lastly, it should be mentioned that the incoming data is complex-valued and that correlations must be done for both the I (in-phase) and Q (quadrature phase) channels, doubling the hardware cost for the accumulation.

9.1.1. Architectural Exploration (0.8μ CMOS design)

As the weighting function for the correlation is simply a sign-toggle, the operation effectively reduces to an accumulation. The main element of functionality is the addition/subtraction of the 3 magnitude bits over 64 samples at 64 MHz, requiring 9 bits of magnitude plus 1 bit of sign for dynamic range, with a resulting output dump frequency at 1 MHz. As described in Chapter 8, the timing recovery and multipath estimators require longer correlation (over 1024 samples), which will require an additional 4 bits, or 14 bits total of dynamic range. These long correlations are achieved by taking the absolute value of the 64-sample correlations, and further accumulating another 16 cycles. As a simple-minded first attempt, one possible implementation is to use a straightforward two's-complement ripple adder and literally accumulate the data after multiplication by the weighting sequence (Figure 9.1). Since the incoming data is sign-magnitude, it would need to be converted to two's complement for the above approach. This sign-extension causes significant additional power [Chandra94], but it is not the worst aspect. The carry chain for a ripple adder must complete in 15.6 ns (64 MHz) minus the register setup and delay times and any clock skew. From the process characterization data for the available 0.8μ technology, it was found this allows for 26 inverter delays, or about 13 gates. Allot-

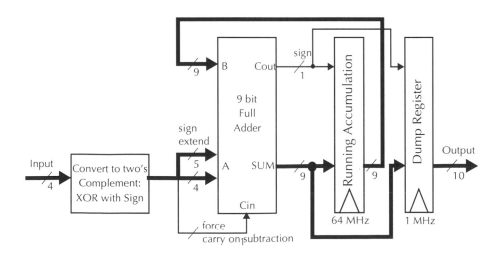

Figure 9.1: Simple 64-sample correlator architecture

ting at least one gate delay per carry operation, there are only 3 gate delays left for the register and carry setup circuitry. It might be possible to design the circuit to operate with this very small amount of overhead, but it is not promising. To be sure, there are several well-known techniques to help speed up this circuit, including running at a higher voltage, carry-select addition, carry look-ahead addition, etc., but due to the tight timing constraints and desire for low power a different approach was examined. Increasing the voltage (above 1.5V) was unacceptable from a power perspective; more complex adders tend to drastically increase in power and area. The inability to utilize a simple ripple adder implies that a rather deeply pipelined scheme will need to be employed.

To preserve the advantage of the sign-magnitude nature of the incoming data, one architectural idea is to immediately break up the accumulation into two parts: an accumulation of all incoming positive data and an accumulation of all incoming negative data. The sign bit can be used to multiplex the 3 bits of magnitude to the either the positive or negative accumulator, and the difference between positive and negative can be computed after dumping at a 1 MHz rate by simply including a subtractor after the dump register. This has the advantage that the final subtractor will take negligible power at 1 MHz and has plenty of time to compute, but increases the area for the correlator slightly. In the following discussion, the positive-data accumulator will be designated POSACC, and the negative-data accumulator will be designated NEGACC (Figure 9.2).

Another architectural optimization is to cut down the critical path by pipelining the entire carry chain to reduce the critical path. The degree of pipelining is arguable; more registers ease the carry path design, but increase clock power and area. To examine this trade-off, the SPICE circuit simulator was used to determine the delay for a latch, a single bitslice[1] of the adder (which consists of two exclusive OR's), and another latch. The latch delay was simulated to be 3.5 ns with a 2 ns setup time at 1.5V, and a single XOR delay around 3.5 ns. Using these numbers, to simply latch, add, and latch again requires 12.5 ns Since the cycle time is 15.6 ns, this implies that the carry would have to be bit-pipelined; a carry-save architecture is the only viable choice at these speeds [Rabaey96]. A carry-save architecture entails the use of two register banks, one to hold the current sum vector and one to hold the current carry vector. The cost of this replication is extra area for registers, and a separate adder to recombine the sum and carry vectors after the 1 MHz dump operation; furthermore, extra power is needed to clock approximately twice as many registers. In spite of the extra complexity, this proved to be significantly less power than resorting to a higher supply voltage and more complex adder structures such as a carry-lookahead [Chandra94]. This is architectural voltage scaling; the requirement for the 1.5V supply rail dictated the architecture of the design. Thus, the bit-level carry pipelining of the carry-save reduces to the critical path to that of a single bitslice of a full adder cell. The choice of a carry save architecture will be revisited later in this chap-

---

1. A bitslice is a single unit of the overall structure. Since this is a 9-bit adder, there are 9 parallel "subunits", each of which is considered a bitslice.

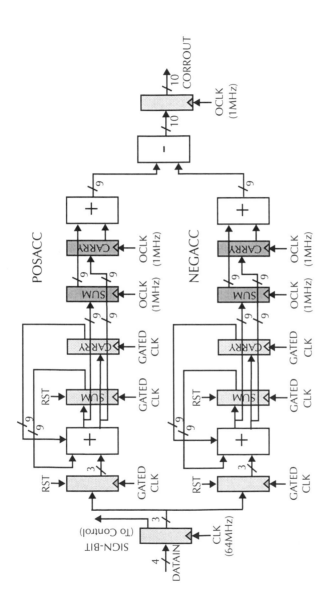

Figure 9.2: Carry-save, sign magnitude correlator architecture [Stone95] POSACC is the positive data accumulator, and NEGACC is the negative data accumulator.

ter for the second version of the correlator, where a more advanced silicon process will allow us to remove extra registers in the carry path. However, the choice of a carry-save based correlator is a sound one for a low-area, low power implementation.

9.1.2. The Carry Save Bit Slice

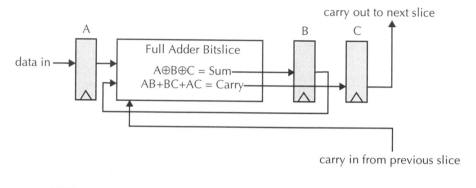

Figure 9.3: Critical path for correlator design

The critical delay path for a carry save bit-slice, as shown in Figure 9.3 above, is given by $T_{clk2Q}+T_{fulladd}+T_{setup}$. The issues that need to be examined are cell design, control, and clocking. True single-phase clocking (TSPC) is used in all of the registers and latches, primarily for speed reasons and for the ease of only having to route one clock line throughout the chip. Likewise, a static CMOS design style was chosen, as it is robust under supply voltage scaling, and can have fewer unnecessary transitions since precharging is not required. As far as transistor sizing is concerned, the aim is to keep the gates as close to minimum size as possible to keep the overall area and power low, up to the constraint set by the critical path.

Not surprisingly, the adder is one of the most studied digital blocks, and an overview of the more common designs may be found in the references [Omondi94] and [Rabaey96]. While there exist many interesting complex CMOS implementations, most are intended for 5V or 3.3V operation, suffering from serious performance degradation at low voltages due to the large number of stacked transistors in the complex gates (The PMOS device rapidly loses drive ability when stacked more than 2 or 3 transistors deep). In addition, more complicated topologies such as carry look-ahead or carry-bypass were eschewed in favor of a simpler, regular structure. For the sum calculation, a three-input exclusive OR function (XOR) will be necessary; the carry generation can be done as a single complex gate or the cascade of several smaller gates. The design decisions for both the sum calculation and the carry generation are driven by the critical delay in the adder block. A complex 3-input XOR gate is too slow and unwieldy; thus, the sum calculation is implemented with two cascaded XOR's instead. As two cascaded XOR's constitutes a longer delay path in the adder that the carry generation, the non-critical carry logic may be implemented in a single, complex gate as opposed to several simple gates. Although the complex gate is slower, the effective capacitance switched is smaller than for a cascade of several smaller gates, resulting in lower power consumption [Rabaey241]. The gates were implemented as shown in Figure 9.4 and Figure 9.5; it should be noted that, for the XOR, the inverse of A and B are delayed, and placed closer to OUT to improve the overall speed of the gate.

The transistor sizing for the XOR gate is chosen by the simple scaling technique commonly used in digital design [Rabaey96]. The NMOS devices are scaled approximately twice from the minimum size allowed by the process as they are stacked two deep. The PMOS are scaled up by a factor of four from the NMOS to equalize the rise and fall times. As the XOR constitutes the critical path, it was oversized for faster operation, while the carry generation gates were kept near the minimum size. Simulation was also performed to verify that the sizes allowed for enough margin (approximately 3 ns) in the critical path. A more aggressive design could further reduce power consumption by sizing down the transistors at the risk of missing the cycle time; this is risky under process and temperature variation, and the power reduction could not be justified in light of this.

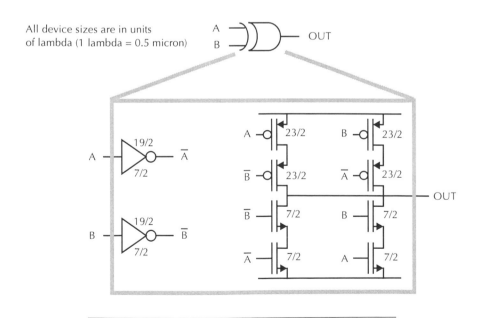

Figure 9.4: XOR gate implementation

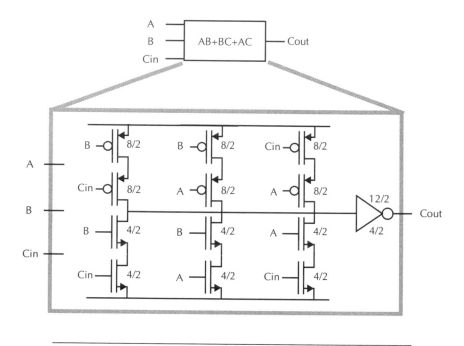

Figure 9.5: Carry generation gate implementation

For the register block, the TSPC register cell is shown in Figure 9.6. The TSPC register is mostly minimum size except for a slightly sized up frontend stage for quicker set-up and a large evaluation NMOS (and consequently a sized up PMOS on the next stage) to speed up the slow path of evaluation through the gate [Burd94, Yuan89].

9.1.3. Performing the Weight Multiplication

The accumulators discussed thus far deal only with the summation of data, thus the multiplication by +/-1 from both the PN and Walsh codes still needs to be performed. As this is a trivial multiply, it is simply the XOR of the incoming data's sign bit with the PN and Walsh bits. As with the carry save adder, the clocking period is only able to allow safely two XOR delays between registers. Fortunately, that is identical to what must happen to perform the weight multiplication, so the same design can be utilized. However, another two pipeline registers will need to be added to the front of the datapath to allow enough time to perform the multiplication and then feed the result to the control logic to multiplex the data to the proper (POSACC or NEGACC) accumulator datapath (Figure 9.7).

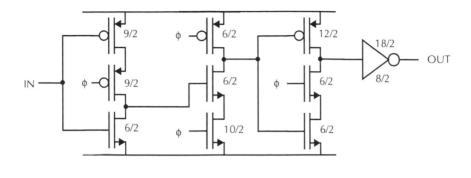

Figure 9.6: TSPC register implementation

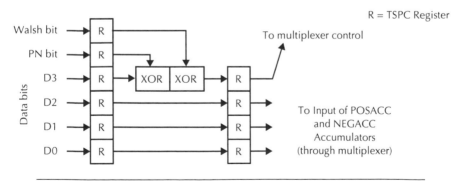

Figure 9.7: Walsh and PN weight multiplication

9.1.4. Correlator Control Signals

There is a minimal amount of control that needs to be designed to perform the multiplexing between the positive and negative accumulators, and to accommodate a reset. Although considered by some to be risky, the technique of gated clocks is used, as it is better from a power consumption standpoint to not clock sections that are not in use. In particular, this reduces the power consumption in the accumulator by almost a factor of two, as only one of either the positive or negative accumulators is clocked per sample. Two control signals are added to the correlator: DUMP (an enable for latching the dump registers and resetting the running accumulation registers), and RESET-DUMP (an enable for resetting both the dump and running accumulation registers). The desired control functionality is (on a rising clock edge):

1. Dump registers take sum and carry vectors on DUMP assertion

2. Dump registers reset on RESET-DUMP assertion

3. POSACC updates running accumulation register for positive data ($\overline{\text{Sign}}$) or DUMP

4. POSACC resets running accumulation registers on DUMP

   (This seems redundant, to update and reset on DUMP, however, the reset in the TSPC registers is an enable, only evaluating to low after a clock edge.)

5. NEGACC updates running accumulation register for negative data (Sign) or DUMP

6. NEGACC resets running accumulation registers on DUMP

7. POSACC input register clocks on $\overline{\text{Sign}}$

8. POSACC input register resets on (DUMP and Sign)

   (Important to not miss the first sample of the next correlation when dumping/resetting)

9. NEGACC input register clocks on Sign

10. NEGACC input register resets on (DUMP and $\overline{\text{Sign}}$)

This is relatively easy to provide, once the sign of the data is known. After the sign bit is known, it is quickly inverted and NOR'ed appropriately to provide the needed control signals before the falling edge of the clock. All control signals are clocked in on the falling edge, to provide a complete half cycle for the clock to be stable before the datapath itself clocks on the rising edge.

9.1.5. Two's Complement to Sign-Magnitude Conversion

At this point, the correlator design is nearly complete. One of the key remaining issues is that of the 1 MHz subtractor that recombines the results of the positive accumulator POSACC and the negative accumulator NEGACC. The subtraction itself is quite simple, especially given that the circuitry is operating at the decimated frequency (post-dump) of

1 MHz. The data from NEGACC is inverted bitwise, and summed with "1" plus the data from POSACC. Effective, the two's complement negation of NEGACC is performed (bitwise inversion plus a sum with "1"), and then added to the POSACC data.

However, the result (POSACC-NEGACC) will then be in two's complement format, as opposed to sign-magnitude. For longer correlations, the absolute value of the signal is required, which may be easily accomplished by simply ignoring the sign bit of a sign-magnitude number. Furthermore, for DQPSK data decoding, magnitude multiplications and combine (add or subtract) afterwards are used to simplify the multiplier design. Even though the power benefits of sign magnitude versus two's complement are small at 1 MHz, there are some strong system issues dictating a sign-magnitude representation at the correlator output instead of two's complement.

Given the low (1 MHz) operating frequencies, the issue becomes one of how to do the conversion in a small amount of area. If the outcome is positive, nothing needs to be done; if the result is negative, the data needs to be negated to derive the magnitude, and the sign-bit set to 1 to indicate a negative value. Unlike the negation described above, this negation is conditional, and there is no adder conveniently available to perform the additional sum-with-one required. In this case, a straightforward method is to run the two's complement output of the correlator into a decrementer, or half subtractor, with the decrement equal to the sign bit (which will subtract 1 from the data if negative, 0 otherwise), and then run the output of that into a bank of XOR's which conditionally bit-wise invert on the sign bit. Effectively, if the sign bit is 0, nothing is subtracted, and no inversion is performed; the opposite is true of the sign bit is 1. Thus, a minimum amount of control is used while achieving the desired functionality. A block diagram is shown in Figure 9.8.

It should be noted that the two's complement to sign magnitude conversion is only one of three functions that happen after dump. In addition to the conversion, the sum and carry vectors of the carry-save architecture needs to be recombined, and the NEGACC and POSACC data needs to be recombined – all of these are performed with simple ripple carry adders with no pipelining. Even with 1000 nsec (1 MHz) to perform, this still may be stretching the timing requirements at a low supply voltage of 1.5V. Fortunately, this is not the case; from SPICE simulations, a set of 3 full ripples of 9 bits requires 100 ns to complete, or 10% of the 1 MHz clock period. Again, the architectural choices are driven by the constraints of low supply voltage and critical delay path. The robustness of static CMOS under voltage scaling allows the use of standard ripple adders at 1.5V while meeting the 1000 nsec delay, making a small-area implementation the key goal.

9.1.6. Clock Buffering

As was mentioned above in Section 9.1.4 above, control is achieved by gating the clock for the correlator. This is a relatively simple scheme where the global clock is gated by a NAND operation, then buffered with an inverter for drive (Figure 9.9). In clocking the datapath registers, the main issue is minimizing skew between register banks, as excessive skew may either adversely impact the critical path, or cause incorrect latching. Another

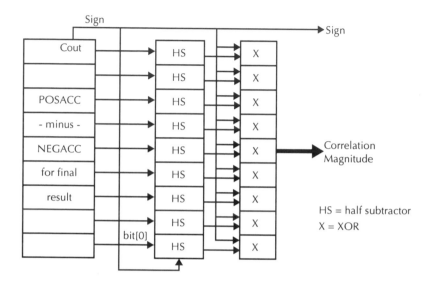

Figure 9.8: Two's complement to sign magnitude conversion

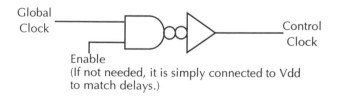

Figure 9.9: Clock gating with a NAND

issue is that of the clock edge rise-fall times, since the TSPC registers are sensitive to long rise-fall times. The inverter buffer is sized to give fast enough edges (from SPICE simulation, rise/fall less 4 nsec). Again, the control is set up by clocking on the falling edge to provide the enable signal several nanoseconds to stabilize before the rising edge of the global clock. These critical timing issues were verified with SPICE by simulating the extracted layout of the entire control section.

One straightforward way to break up the clock load and to ensure minimum skew is to match or balance the capacitive load seen by the inverter buffer. This can attained by grouping the registers into banks of approximately the same size. For example, the sum registers (9 bits) and carry registers (8 bits) can be placed in separate banks. The input has 8 bits (4 data bits, PN, Walsh, DUMP, RESET-DUMP), and may be considered to be a bank also. The remaining registers are: intermediate control registers (clocked by the falling edge), and the 3-bit input registers to the accumulators. Since the input registers have a third of the bits, their drivers can be reduced in size to compensate. Likewise, the intermediate registers (two banks of 6 bits) may be driven by an inverter two-thirds the size of the default inverter-buffer cell.

With this as a guide to the relative scaling of the inverter buffers in each bank, the capacitive clock load of a single register needs to be determined. From process characterization data for the 0.8 micron process, the gate capacitance of the register can be determined from the transistor sizings described in Section 9.1.2. Table 9.1 gives these values for various number of registers, along with an estimate of the required buffer sizing. The factor of 2 in the C_{load} estimate is to account for the extra capacitance in the metal wiring, as well as the source/drain parasitic capacitance of the buffer itself. The W/L sizing estimates are derived from I-V characterization data for the NMOS devices in the 0.8 micron process; it is the size (in width-to-length ratio) of the transistor necessary to generate the required current I at a gate voltage of 1.5V (the supply voltage).

Clocking Load	Cload Est.	I for 1.5V in 4 ns	Power @ 64 MHz	NMOS W/L Est.
4 registers	2*(142.8) fF	113 µA	22 µW	11 (22λ/2λ)
8 registers	2*(285.6) fF	215 µA	43 µW	22 (44λ/2λ)
16 registers	2*(571.2) fF	428 µA	85 µW	44 (88λ/2λ)
32 registers	2*(1.142) pF	857 µA	170 µW	88 (176λ/2λ)

Table 9.1: W/L Estimates for Clock Drivers.
I is the current required to slew C_{load} from 0 to 1.5V in 4 nsec;
W/L is the sizing estimate from process data to generate that current level
(at a gate voltage of 1.5V).

The actual sizing chosen for the inverter buffer was 10.2μ/.8μ for NMOS device. The PMOS increased by approximately 3 times over the NMOS to 32μ/.8μ for driving 9 registers. The ratio of 3 is to account for some of the difference in current drive capability between the NMOS and PMOS devices. To truly match drive capability (and hence create equal rise-fall times in the logic), the PMOS devices would need to be 4 to 5 times larger, requiring much more area and increased power to drive the parasitics. A ratio of 3 is a good compromise between reasonable matching of rise-fall edge rates and maintaining small area and low power consumption. Using the SPICE circuit simulator, the edge rates have been verified at 4.5 ns, with a propagation delay of 3 ns. This verification is done by two methods. First, the NAND gates are simulated driving an inverter-buffer loaded with the estimated capacitance for the appropriate number of registers to check our size estimates. Secondly, the transistors and clock lines are extracted from the actual, completed correlator layout and performance verified again with SPICE.

Obviously, the key question for the clock tree is, "How much skew can really be tolerated?" The answer may be analyzed for cases of both positive negative skew. On one side, if the clock arrives at registers further in the datapath sooner than in previous ones, this results in a penalty for the critical path. Recall from Section 9.1.1 that there is approximately 3 ns of margin in the critical path, which should be much more than any skew with careful load balancing. If the clock edge to the earlier registers occurs before that of the later registers, a race-condition may occur, where the new data overwrites the old before it has a chance to be latched into the register. In the correlator, the smallest delay path is a pair of back-to-back registers with no logic delay between them for the three input magnitude data bits. The clock edge to output delay in the TSPC latch is approximately 3.5 ns, implying that the most skew could be tolerated is half that, or 1.7 ns, to allow for some safety margin. Thus, as long as the skew between any two registers is less than 1.7 ns, no skew-induced data errors should occur. By using such load balancing techniques, simulation results confirm that the skew due to loading is less than 1 ns.

9.1.7. Power Estimation for the Correlator

In general, power estimation for digital circuitry is done by running a random set of vectors through the logic and having a program count the amount of switched capacitance per cycle, thus determining an average switching capacitance C_{eff}. Power is then determined from the equation $C_{eff}V_{dd}^2 f$, since V_{dd} and f are already known. In addition to pure simulation, hand-estimates can be performed to verify that the simulation results are roughly correct. Taking power as having two components: power to the clock and power of the data moving through the logic, the overall power for the correlator block can be estimated by making some basic assumptions. Since the circuit is bit-pipelined with only a couple of gates between registers, we can assume that the power of the data will be roughly equal to the power of the clocking as they have roughly equal logic depth. The clock load for an accumulator (40 registers with approximately 40 fF load each) is 1.6 pF. If we assume this is charged linearly in 4 ns, this implies a current of CV/Δt = 0.6 mA. This is occurring at a 64 MHz rate, so the power to drive the clock is I*(4/15.6)*1.5V=0.23 mW. Using our bold assumption that logic power equals clocking power, we can double that power to 0.46 mW. Also, as the clock has to drive its own load in addition to the gates of the registers, we may assume that the clock network's load is roughly equal to the gate load, so add another 0.23 mW for a total of 0.7 mW for a corr-elator. Since it is a complex data stream, there are two correlators in each complex corre-lation, yielding 1.4 mW for I and Q correlations.

To achieve a more accurate power estimate, the actual layout can be extracted and used in a simulator to determine effective switching capacitance, and hence power consumption. In our case, the IRSIM-CAP simulator [Landman95] was used, with spread-spectrum-like random input data. For the 0.8μ process, IRSIM-CAP predicts a power consumption of 0.66 mW, or 1.32 mW for I and Q. Finally, to compare with an actual measurement, a single test correlator was measured to have a power of 0.6mW at 64 MHz (1.2 mW for a complex correlation) which agrees well with simulation.

The final layout of both the 64-length (front-end) correlator and the 1024-length (long) correlator are shown in Figure 9.10. They are both compact, with little wasted silicon ar-ea, and implement the correlation operation using the carry-save architecture described in the preceding sections.

The final sizes of the correlators are:

Basic correlator (64 samples):
In the 0.8μ process = 375μ tall x 1100μ wide = 0.41 mm^2 (0.82 mm^2 for I+Q)

Long Correlator (1024 samples):
In the 0.8μ process = 375μ tall x 1330μ wide = 0.5 mm^2 (1.0 mm^2 for I+Q)

(a)

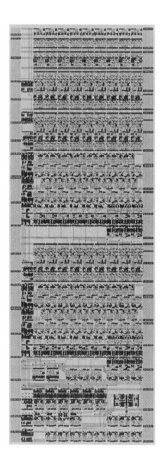

(b)

Figure 9.10: Final layout for correlators.
(a) Long (1024-length) correlator
(b) Basic (64-length) correlator

REFERENCES

[Burd94]   T. Burd. *Low-Power CMOS Library Design Methodologies*. M.S. Thesis, U.C. Berkeley, Berkeley, CA, June 1994.

[Chandra92]   A. Chandrakasan, S. Sheng, and R.W. Brodersen. "Low-Power CMOS Digital Design." *IEEE Journal of Solid-State Circuits*, Vol. 27, No. 4, pp. 473-484. April 1992.

[Chandra94]   A. Chandrakasan. *Low Power Digital CMOS Design*. Ph.D. Thesis, U.C. Berkeley, Berkeley, CA, 1994.

[Landman95]   P. Landman. *Low-Power Architectural Design Methodologies*. Ph.D. Thesis, U.C. Berkeley, Berkeley, CA, August 1994.

[Omondi95]   A. Omondi. *Computer Arithmetic Systems: Algorithms, Architecture, and Implementations*. New York: Prentice-Hall Inc., 1994.

[Rabaey241]   J. Rabaey. EECS 241 Digital Circuit Design Class Notes. U.C. Berkeley, Berkeley, CA, Spring 1996.

[Rabaey96]   J. Rabaey. *Digital Integrated Circuits: A Design Perspective*. New Jersey: Prentice Hall Inc., 1996.

[Stone95]   K. Stone. *Low Power Spread Spectrum Demodulator for Wideband Wireless Communications*. M.S. Thesis, U.C. Berkeley, Berkeley, CA, 1995.

[Yuan89]   J. Yuan, C. Svensson. "High-Speed CMOS Circuit Techniques." *IEEE Journal of Solid-State Circuits*, Vol. 24, No. 1. February 1989.

10 Conclusions and Future Directions

In providing future personal communication services, broadband multimedia data services - the most demanding of which being real-time, on-demand video - will be delivered over wireless transmission systems to high densities of mobile users. Even with the best video and data compression schemes available today, this would require a per-user sustained data rate of from 1-2 megabits per second – a sustained rate far higher than what existing cellular phones can provide and, because of the real-time requirements, far higher than available from existing wireless LAN's. This book has been an investigation into a low-power CMOS implementation of significant portions of a broadband downlink (from the base station to portable) that meet these requirements, with particular emphasis on design techniques to provide the lowest possible power consumption of the mobile receiver functions while providing high-performance data throughput.

The use of spread spectrum techniques, in particular direct sequence code-division multiple access (CDMA), provides the multiple-access strategy to maintain parallel, separate streams of real-time data to all the users. Since the base station can provide synchronization and serves as a single point of transmission, the near-far problems and capacity issues with traditional direct-sequence systems can be avoided. Likewise, the intrinsic noise rejection inherent in spread-spectrum is particularly desirable, as it can be used to reduce sensitivity to interference and narrow band multipath fades in the radio environment, as well as relax hardware performance constraints.

Since the overall goal is to provide high-speed wireless communications in a low-cost, portable device, a standard digital single-poly 0.8 micron CMOS technology is used in the implementation of both the analog and digital circuitry. Three custom integrated circuits have been designed which implement most of the transceiver path: the digital baseband modulator; the analog receiver front-end and A/D; and the digital receiver baseband processor. The techniques used in their design have been described in the text.

The transmit architecture of the system is based loosely on that of the IS-95 digital cellular standard, scaled to a spread-spectrum chipping rate of 64 MHz; it uses a hybrid Walsh-PN code to achieve the desired cross-correlation and autocorrelation properties. To aid receiver synchronization, the zeroth Walsh code is expended as a pilot tone. This minimizes the probability of false lock, and enables the receiver to perform continuous multipath profile monitoring as well as adjacent cell detection. A monolithic prototype modulator has been developed for the base station, implementing all of the baseband processing required before single-sideband upconversion to the RF carrier; this includes spreading each data bit with a unique 64-long Walsh code, and then modulating the combined user data streams with the 32768-long PN code. To provide better control of capacity and interference within each cell, the separate data streams are combined with provision for individual user power control, as well as an overall power scaling factor. Lastly, to bandlimit the transmit signal in a controlled, zero-intersymbol interference fashion, a digital raised-cosine filter has also been included on the modulator prototype; this filter effectively runs at 256 MHz, implemented as four interleaved 64 MHz subfilters. Currently, the die accepts 15 parallel user inputs, and several die can be synchronously connected to achieve up to 50 users operating from a single base station. Lastly, for analog upconversion purposes, a semicustom single-sideband RF modulator has been implemented, providing digital-to-analog conversion of the prototype modulator's output and frequency translation to an RF carrier of 1.088 GHz.

Clearly, the most critical design issues of the hardware lie in the development of the receiver radio-frequency components. In addition to the requirement of low power consumption, the analog RF circuitry is complicated by the need to operate at frequencies near or above 1 GHz. Such circuitry has been traditionally dominated by designs using discrete GaAs, silicon bipolar transistors and stripline filters, which consume significant area on a circuit board and excessive amounts of power, especially when matching to standard 50Ω impedance levels. However, by the use of proper tuning networks, CMOS operation in the 1 GHz band has been demonstrated to be quite feasible, even when using only a 0.8 micron standard digital CMOS technology. The continued scaling of this technology will allow ever higher frequency ranges to be achieved.

At the receiver, the incoming RF signal is amplified by a continuous-time low-noise amplifier, and is mixed to near baseband (actually 32 MHz) via direct sampling of the modulated signal. This eliminates the need for a conventional superheterodyne mixing and IF processing, and makes it possible for the variable gain function of the AGC to be implemented in sampled-data amplifiers. The signal is then A/D converted using dual interleaved 4-bit flash converters at 128 MHz (64 MHz I and Q). The remainder of the processing - timing and data recovery - is executed by the digital baseband receiver circuit, which performs coarse and fine timing recovery using scanning correlators and a delay-locked loop with a time resolution of 8 nsec. After timing acquisition is obtained, the I and Q data are then despread using length-64 matched-filter correlators. As a demonstration vehicle, a testchip has been built that implements this functionality at half-speed, a clock rate of 32 MHz, limited only by the available 0.8 micron fabrication technology.

A two-chip prototype receiver - consisting of a monolithic analog front-end and a high-speed baseband DSP - has been built and tested running at a carrier of 1.056 GHz. The total peak power consumption for the receiver prototype (both analog and digital integrated circuits) is 154 mW, far lower than any receiver reported to date, and especially so in light of the supported data rate.

The large power savings that has been demonstrated arise from the architectural design of the receiver: the chosen architecture is quite unlike the superheterodyne architectures used in radio systems to date. Channel selection, timing recovery, data recovery, and frequency recovery are all performed in the digital domain; the analog front-end consists only of amplification, filtering, and a fast, low-resolution A/D conversion. The reason for this is clear: low-power, high-speed digital signal processing can be achieved through architectural voltage scaling techniques in the digital domain – techniques that do not translate easily into the analog domain. In spite of the tremendous amount of digital processing required, and the minimization of the amount of analog hardware required by the system, the power consumed by the prototype receiver DSP is still less than one half of that consumed by the analog front-end – testament to the effectiveness of these techniques. Such an "A-to-D front-end" receiver architecture hinges on this: by pushing the analog/digital boundary as close to the antenna as possible, these digital techniques can be brought to bear early in the signal chain, resulting in a receiver that can process a broadband multimedia data stream while consuming a total of 154 mW of power.

With this in mind, several directions can be taken for future research. First, the limits of tuned-CMOS RF can be explored further: there is freely available, unlicensed spectrum in the 2 GHz and 5 GHz bands. Given recent research in on-chip inductor tuning techniques [Ou95], one potential avenue of research is the adaptation of sampling demodulation techniques to these higher frequencies. While this project has demonstrated frequency limits due to jitter of approximately 1 GHz for a 0.8 micron process, 0.5 micron and 0.35 micron processes are becoming readily available, with concomitant improvements in sampling-edge jitter performance.

Furthermore, it is not clear that the RAKE strategy is optimal in the case of correlated interference, as what is seen in an indoor environment. Recent research [Teus97] has shown that there may be better multipath combining strategies than conventional RAKE; likewise, application of trellis codes and Viterbi detection may also yield gains at the system level, especially with regard to relaxing analog performance requirements. Along these lines, with the high levels of integration and low power consumption demonstrated, multiple receivers with phase-coherent combining become quite viable in an integrated implementation. Effectively, a monolithic, electronically steerable antenna array could be formed, further improving multipath rejection and improving SNR at the receiver.

Lastly, there is a great deal of current interest with regard to discrete multitone (DMT), also known as orthogonal frequency-division multiplexing (OFDM) [Ho96, Zog95, Wu95]. In some sense, OFDM provides spreading redundancy in the frequency domain: the time domain and the frequency domain are interchanged through the use of a Fast

Fourier Transform in both the transmitter and receiver. DMT-based systems have become popular in wired communications, especially in emerging broadband copper services like the asymmetric digital subscriber loop (ADSL) [ADSL95]. DMT promises the benefits of spread-spectrum, while improving immunity to multipath and narrowband interference. However, it requires precise knowledge of the transmission channel profile: adaptation into a wireless environment, where fading is a serious issue, is an open research question. In any case, a wireless DMT system can leverage off the design techniques explored in this project – compensating for the performance limitations of integrated CMOS RF analog by careful design tradeoffs at the system level, and by exploiting low-power, high-performance digital techniques as much as possible.

REFERENCES

[ADSL95]   *ADSL Metallic Interface Standard.* ANSI Document T1E1.4/95-007R2/T1.413. American National Standards Institute, 1995.

[Ho96]   M. Ho M, J.M. Cioffi, and J.A.C. Bingham. "Discrete Multitone Echo Cancellation." *IEEE Transactions on Communications*, Vol. 44, No. 7, pp. 817-825. July 1996.

[Ou95]   J.J. Ou. "CMOS Low-Noise Amplifier Design for Monolithic RF Modems." Ph.D. Qualifying Exam Presentation, U.C. Berkeley, Berkeley, CA, November 1995.

[Teus97]   C. Teuscher. Ph.D. thesis on Multi-Access Limited, Multi-User Combining Techniques, Forthcoming. U.C. Berkeley, Berkeley, CA, 1997.

[Wu95]   Y.Y. Wu and W.Y. Zou. "Orthogonal Frequency Division Multiplexing- A Multi-Carrier Modulation Scheme." *IEEE Transactions on Consumer Electronics*, Vol. 41, No. 3, pp. 392-399. August 1995.

[Zog95]   T. Zogakis, J. Aslanis, and J.M. Cioffi. "A Coded and Shaped Discrete Multitone System." *IEEE Transactions on Communications*, Vol. 43, No. 12, pp. 2941-2949. Dec. 1995.

Index